Indiana
HF
181-TFG
AF
67 384

F-4 PHANTOM IIs

OF THE USAF RESERVE AND AIR NATIONAL GUARD

Also by the Author

USAF F-15 Eagles: Units, Colors & Markings

Republic's A-10 Thunderbolt II: A Pictorial History

The Boeing C-135 Series: Stratotanker, Stratolifter and other Variants

Northrop's YF-17 Cobra: A Pictorial History

General Dynamics F-111 Aardvark

Northrop's T-38 Talon: A Pictorial History

Rockwell B-1B: SAC's Last Bomber

ACC Bomber Triad: The B-52s, B-1s, and B-2s of Air Combat Command

The 388th Tactical Fighter Wing at Korat Royal Thai Air Force Base 1972

F-4 PHANTOM IIs
OF THE USAF RESERVE AND AIR NATIONAL GUARD

Don Logan

Schiffer Military History
Atglen, PA

ACKNOWLEDGEMENTS

I would like to thank Brian C. Rogers for his help in the research for this book and to Tom Brewer and William Peake for MiG Kill and other serial number specific information. Thanks to Craig Brown and Jerry Geer for the unit patches used in this book.

Special Thanks to Don Spering/Aircraft in Review (AIR) for the special event markings designed by him, which were added to many ANG, and reserve Phantoms. He has also provided many photos of these specially marked phantoms for this book.

Thanks to Doug Barbier for the photos in the 191st FIG, Michigan Air National Guard chapter of this book.

I would also like to thank the following individuals who provided photographs included in this book: Mark Bacon, Doug Barbier, Tom Brewer, David F. Brown, Jack Callaway, George Cockle, Bill Curry, Robert F. Dorr, D.J. Fisher, Alec Fushi, Maj General Wayne Gatlin, Mike Grove, Jerry Geer, James Goodall, Norris Graser, Bob Greby, Mike Grove, Phillip Huston, Marty Isham, Kevin Jackson, Craig Kaston, Ben Knowles, Ray L. Leader, Bob Leavitt, Geoff Lebaron, Nate Leong, Don Linn, Frank MacSorley, Charles B. Mayer, Donald S. McGarry, Frank McCurdy, J.E. Michaels, Robert J. Mills, Jr., Kirk Minert, Paul Minert, Jack Morris, Kevin Patrick, William R. Peake, Robert J. Pickett, Paul Osborne, Doug Remington, Charles T. Robbins, Brian C. Rogers, Barry Roop, Jim Rotramel, Bob Shane, Douglas E. Slowiak/Vortex Photo Graphics, Keith Snyder, Don Spering/AIR, Bill Spidle, Bruce Stewart, Jim Tunney, Pete Wilson, Scott Wilson, Scott Van Aken, Ted Van Geffen, and McDonnell Douglas/Boeing.

Book design by Robert Biondi.

Library of Congress Catalog Number: 2002104066.

Printed in China.
ISBN: 0-7643-1627-3

We are always looking for people to write books on new and related subjects. If you have an idea for a book, please contact us at the address below.

Published by Schiffer Publishing Ltd.
4880 Lower Valley Road
Atglen, PA 19310
Phone: (610) 593-1777
FAX: (610) 593-2002
E-mail: Schifferbk@aol.com.
Visit our web site at: www.schifferbooks.com
Please write for a free catalog.
This book may be purchased from the publisher.
Please include $3.95 postage.
Try your bookstore first.

In Europe, Schiffer books are distributed by:
Bushwood Books
6 Marksbury Ave.
Kew Gardens
Surrey TW9 4JF
England
Phone: 44 (0)208 392-8585
FAX: 44 (0)208 392-9876
E-mail: Bushwd@aol.com.
Free postage in the UK. Europe: air mail at cost.
Try your bookstore first.

Contents

Introduction 7

AIR FORCE RESERVE UNITS 8

93rd TFS, 482nd TFW, Homestead AFB, Florida 8

465th TFS, 507th TFG, Tinker AFB, Oklahoma 15

89th TFS 906th TFG, Wright-Patterson AFB, Ohio 24

704th TFS, 924th TFG, Bergstrom AFB, Texas 30

457th TFS, 301st TFW, Carswell AFB, Texas 39

AIR NATIONAL GUARD UNITS 48

106th TRS, 117th TRW, Birmingham Municipal Airport, Alabama 49

160th TRS/160th TFS, 187th TRG/187th TFG, Dannelly Field, Montgomery, Alabama 75

184th TFS, 188th TFG, Fort Smith Municipal Airport, Arkansas 80

194th FIS, 144th FIW, Fresno ANGB, California 87

196th TFS/196th TRS, 163rd TFG/163rd TRG, March AFB, California 92

121st TFS, 113th TFW, District of Columbia ANG, Andrews AFB, Maryland 102

128th TFS, 116th TFW, Dobbins AFB, Georgia 110

199th TFS, 154th Composite Group, Hickam AFB, Hawaii 114

190th TRS/FS, 124th TRG/FG, Gowen Field, Boise, Idaho 120

170th TFS, 183rd TFG, Capital Airport, Springfield, Illinois 132

163rd TFS, 122nd TFW, Baer Field, Fort Wayne, Indiana 140

113th TFS, 181st TFG, Hulman Field, Terre Haute, Indiana 149

127th TFS and 177th TFTS, 184th TFG, McConnell AFB, Wichita, Kansas 157

165th TRS, 123rd TRW, Standiford Field, Louisville, Kentucky 170

122nd TFS, 159th TFG, NAS New Orleans, Louisiana 176

171st FIS, 191st FIG, Selfridge Air Nation Guard Base, Michigan 181

179th TRS/179th FIS, 148th TRG/148th FIG, Duluth International Airport, Minnesota 192

153rd TRS, 186th TRG, Key Field, Meridian, Mississippi 198

110th TFS, 131st TFW, Lambert Field, St. Louis, Missouri 205

173rd TRS, 155th TRG, Lincoln Municipal Airport, Lincoln, Nebraska 214

192nd TRS, 152nd TRG, Reno-Cannon International Airport, Reno, Nevada 224

141st TFS, 108th TFW, McGuire AFB, New Jersey 231

136th FIS, 107th FIG, Niagara Falls International Airport, New York 245
178th FIS, 119th FIG, Hector International Airport, Fargo, North Dakota 252
114th TFTS, 142nd FIG, Kingsley Field, Klamath Falls, Oregon 258
123rd FIS, 142nd FIG, Portland International Airport, Oregon 264
111th FIS, 147th FIG, Ellington Air National Guard Station, Houston, Texas 269
182nd TFS, 149th TFG, Kelly AFB, San Antonio, Texas 274
134th TFS, 158th TFG, Burlington, Air National Guard Station, Vermont 280

PAINT SCHEMES 284

NAVIGATION AND WEAPONS DELIVERY SYSTEM (NWDS) AN/ASQ-203 288

F-4 SERIAL NUMBERS 292

F-4 MAIN DIFFERENCES 296

AIR FORCE RESERVE & AIR NATIONAL GUARD F-4 LOSSES 297

Introduction

First ordered by the Navy in 1954 as a carrier based fighter interceptor, the F-4 became numerically the most important U.S. military aircraft of the 1960s and early 1970s. Between May 27, 1958, when the first production model of the Phantom II, the F4H made its first flight, and October 26, 1979, when the last American-built F-4 an F-4E (78-0744) was delivered to the Republic of Korea Air Force, a total of 5,068 Phantom IIs were built by McDonnell Aircraft for the United States and allied nations. Fighter versions built for the USAF were the F-4C first delivered to Tactical Air Command in November 1963, F-4D, and F-4E. The USAF reconnaissance version of the Phantom II, the RF-4C first flew in May 1964.

In 1972, as a result of the implementation of the Total Force Policy, which gave reserve components a larger role in National Defense, the Air Force Reserve (AFRES) began organizing Fighter Squadrons. Initial aircraft assigned were F-105D/Fs and A-37Bs. Phantom IIs were first assigned to the AFRES in 1978 when the 93d Tactical Fighter Squadron was activated at Homestead AFB, Florida, with F-4Cs. F-4C, F-4D, and F-4E aircraft were assigned to AFRES. There were no AFRES units equipped with the RF-4C. The last AFRES F-4s were the F-4Es assigned to the two Texas units, the 301st Tactical Fighter Wing (TFW) and the 924th Tactical Fighter Group (TFG), and were replaced by F-16s in 1991.

The first Phantom IIs entering Air National Guard (ANG) service were RF-4Cs. This occurred in February 1971 when the 106th Tactical Reconnaissance Squadron (TRS), 117th Tactical Reconnaissance Wing (TRW), Alabama ANG, began its conversion from Republic RF-84Fs to RF-4Cs. RF-4Cs equipped a total of nine Tactical Reconnaissance Squadrons in the Alabama, California, Idaho, Kentucky, Minnesota, Mississippi, Nebraska, and Nevada Air National Guards.

The first ANG squadron to fly F-4s in the tactical fighter role was the 170th Tactical Fighter Squadron (TFS), 183d Tactical Fighter Group (TFG), Illinois ANG, which received its first F-4Cs in January 1972. Seventeen fighter squadrons (including one TFTS – the 177th at McConnell AFB) in the Alabama, Arkansas, California, District of Columbia, Georgia, Hawaii, Idaho, Illinois, Indiana, Kansas, Louisiana, Missouri, New Jersey, Texas, and Vermont Air National Guards flew all four USAF F-4 Fighter versions: F-4C, F-4D, F-4E, and F-4G.

The F-4G was the last Phantom II version remaining in ANG service, having entered service with the 124th FG, Idaho ANG in October 1991 and retiring in April of 1996.

Between 1978 and 1990, F-4Cs and F-4Ds were also flown in the air defense role. Seven Fighter Interceptor Squadrons (FIS) and one air defense training squadron (the 114th TFTS) in the California, Michigan, Minnesota, North Dakota, New York, Oregon, and Texas Air National Guards flew Phantom IIs. All of these squadrons converted to F-16A/B ADFs or F-15A/Bs.

Individual Units in the Air Force Reserve Section and Air National Guard Section of this book are organized alphabetically by State location. The aircraft serial numbers listed with the Units were all assigned to those Units. There may have been some additional aircraft assigned whose serial numbers were not in my source data.

Air Force Reserve Units

Four AFRES squadrons were equipped with Phantoms during the 1980s, with the last two squadrons retaining their F-4Es until 1991 when they converted to F-16 Fighting Falcons.

	F-4C	F-4D	F-4E
89th TFS, 906th TFG		82-89	
93rd TFS, 915th TFG	78-81		
93rd TFS, 482nd TFW	81-83	83-90	
457th TFS, 301st TFW		80-87	87-91
465th TFS, 507th TFG		80-88	
704th TFS, 924th TFG		81-89	89-91

93rd TFS, 482nd TFW, Homestead AFB, Florida

The first AFRES squadron to fly Phantom IIs, the 93rd TFS came into being in February 1978 when the 915th Airborne Early Warning and Control Group, the parent group of the 79th AEWCS flying EC-121s from Homestead AFB, Florida, was reorganized and redesignated 915th Tactical Fighter Group. In October 1978, the 93rd TFS began flying its F-4Cs carrying **FM** tail code (for **F**lorida **M**akos). Most of the F-4Cs had been transferred from the 401st TFW, Torrejon AB, Spain with a few coming from the 56th TFW at Luke AFB.

Less than three years later, on Apri1 1, 1981, the 915th TFG was inactivated and replaced by the 482nd TFW. The squadron converted to F-4Ds in fall 1983, once again receiving them from the 401st TFW at Torrejon AB. A number of the squadron's F-4Cs were reassigned to the Oregon ANG. The F-4Ds of the 482nd TFW were exchanged for F-16A/Bs in November 1989.

482 TFW AFRES

F-4C (Received from Torrejon AB and Luke AFB)

63-7413	63-7479	63-7484	63-7500	63-7507	63-7532
63-7537	63-7711	64-0655	64-0754	64-0775	64-0796
64-0804	64-0829**	64-0888	64-0890	64-0891	64-0915
64-0922	64-0923				

** 64-0829 - two - MiG-17 - May 20, 1967

F-4D (Received from the 401 TFW, Torrejon AB)

65-0682	65-0781	66-7527	66-7551	66-7552	66-7558
66-7563	66-7609	66-7615	66-7669	66-7679	66-7692
66-7710	66-7731	66-7738	66-7751	66-7767	66-8715
66-8753	66-8804	66-8824			

F-4C 64-0829, a double MiG killer is seen here at Homestead AFB in April 1982 marked as the 482 TFW Flagship with command stripes of the 93rd TFS at Homestead AFB (Black & White), 89th TFS at Wright-Patterson AFB, Ohio (Orange & White) and 704th TFS at Bergstrom AFB, Texas (Blue & White). At the time of the photo these three Air Force Reserve units were under the control of the 482nd TFW. The FM tail code stands for the Florida Makos. 829 is presently painted back in its 8th TFW markings and is on display at the USAF Museum, Wright-Patterson AFB, Ohio. (James Goodall)

F-4D 66-7551 is seen here at Homestead AFB on March 9, 1984 in Europe 1 Camouflage scheme with command stripes of the 93rd TFS at Homestead (Black & White), 89th TFS at Wright-Patterson AFB, (Orange & Black). (Ray Leader)

F-4D 66-7551 was wearing a special three gray test Camouflage scheme when photographed at Homestead AFB on March 9, 1984. It later became a target on the Holloman AFB test range. (Frank McCurdy)

F-4C 63-7537 is seen here at Homestead AFB on May 17, 1983 in an Air Defense gray paint scheme with a small FM tail code. (Ray Leader)

F-4C 63-7507 is seen here at Homestead AFB on May 25, 1980 in the standard Southeast Asia Camouflage scheme. (Ray Leader)

F-4C 64-0923 is seen here at Homestead AFB on August 2, 1981 in the standard Southeast Asia Camouflage scheme. The 93rd TFS Black & White checkered tail stripe and Squadron Insignia are visible on the left side of the aircraft. The aircraft was later transferred to the 163rd TFG, California ANG, and was written off on March 21, 1987. (Pete Wilson)

F-4C 63-7484 was photographed in the standard Southeast Asia Camouflage scheme on July 14, 1980 while visiting Richards-Gebaur AFB. (Don Logan)

F-4D 66-7731 is seen here at Homestead AFB on April 5, 1984 in the Southeast Asia Wraparound Camouflage scheme. (Brian C. Rogers)

F-4D 66-7669 is seen here at Homestead AFB on April 5, 1984 in the Wraparound Southeast Asia Camouflage scheme. (Brian C. Rogers)

F-4D 66-7751 is seen here at Homestead AFB on April 5, 1984 in the Wraparound Southeast Asia Camouflage scheme. (Brian C. Rogers)

F-4D 66-7558 is seen here at Homestead AFB on April 5, 1984 in the Europe I Camouflage scheme. It later became a target on the Holloman AFB test range. (James Goodall)

F-4D 66-7692 is seen here at Homestead AFB on April 5, 1984 in the Europe I Camouflage scheme. On this date, the 93rd TFS had a mix of Southeast Asia Wraparound and Europe I Camouflage schemes. (Brian C. Rogers)

F-4D 66-7731 is seen here at Carswell AFB on March 18, 1987 in the Europe I Camouflage scheme. (Brian C. Rogers)

F-4D 66-7767 is seen here at Homestead AFB in December 1986 in the Hill Grays Camouflage scheme. This aircraft has a non-standard serial number presentation of 67767. (James Goodall)

F-4D 66-7551 is seen here at Homestead AFB in May 1989 in the Hill Grays Camouflage scheme. (David F. Brown Collection)

F-4D 66-7731 is seen here at Homestead AFB in May 1989 in the Hill Grays Camouflage scheme. (Don Linn)

F-4D 65-0682 is seen here at Homestead AFB on October 10, 1987 in the Hill Grays Camouflage scheme. (David F. Brown)

465th TFS, 507th TFG, Tinker AFB, Oklahoma

Activated at Tinker AFB, Oklahoma in May 1972, the 465th TFS was initially equipped with UC tail coded F-105D/Fs. In July 1972 this new squadron was assigned to the 507th Tactical Fighter Group at Tinker. The 465th was next assigned directly to the Carswell AFB, Texas based 301st TFW beginning in March 1973 until the 507th TFG was again activated on October 17, 1975. Following the reactivation, the 465th changed its tail code from UC to **SH** (for **S**ierra **H**otel, phonetic for **S**hit **H**ot, slang for outstanding).

The 465th transferred the F-105s to the AFRES 466th TFS at Hill AFB and converted to F-4Ds in the fall of 1980, receiving the Phantoms from the 474th TFW at Nellis AFB, the 18th TFW at Kadena AB, and the 497th TFS at Taegu AB. The 465th TFS switched to F-16A/Bs during 1988 and was redesignated as a Fighter Squadron in March 1992. The 465th Fighter Squadron received KC-135Rs and was redesignated as the 465th Air Refueling Squadron on April 1, 1994 ending 22 years as a fighter unit.

507 TFG AFRES

F-4D (Received from Nellis AFB, Kadena AB and Taegu AB)

65-0716	66-0272	66-7507	66-7511	66-7515	66-7518
66-7537	66-7548	66-7550★	66-7555	66-7577	66-7608
66-7625	66-7618	66-7673	66-7704	66-7709	66-7715
66-7724	66-7729	66-7732	66-7737	66-7750★	66-7753
66-7758	66-7760	66-7762	66-7773	66-8701	66-8714★
66-8719★	66-8727	66-8735	66-8782	66-8786	66-8816
66-8823#					

★ 66-7550 - MiG 21 - April 16, 1972
★ 66-7750 - MiG-21 - October 24, 1967
★ 66-8714 - MiG-21 - February 5, 1968
★ 66-8719 - MiG-17 - December 17, 1967
66-8823 Destroyed in a mishap on August 21, 1989

Below: F-4D 66-7773 marked as the 465 TFS Flagship escorted by F-16A 80 0507 marked as the 507th TFG Flagship flies past the 507th TFG Ramp at Tinker AFB. (507th Public Affairs Office)

F-4Ds 66-7511 and MiG killer 66-8714 are escorted by F-16A 80-0507 marked as the 507th TFG. (507th Public Affairs Office)

A line of 507th F-4Ds are seen here on the Nellis Flight Line on October 15, 1985 while taking part in Gunsmoke 85. (Marty J. Isham)

F-4D 66-7704 is seen here on May 21, 1981. It is painted with the double blue fuselage stripe which designates the aircraft as the 507th TFG commander's aircraft. (Tom Brewer Collection)

F-4D 66-7555 is seen here on the ramp in April 1981, marked with the single blue fuselage stripe which designates the aircraft as the 465th TFS commander's aircraft. Members of the fire department wash down the ramp after a JP-4 fuel spill. 7555 was sold to the Republic of Korea under the Peace Pheasant program. (Don Logan Collection)

F-4D 66-7507 is seen here on the ramp marked as the 507th TFG Flagship on June 11, 1987. The SH tail code has been attributed to many things from Sooner Home to the politically incorrect Shit Hot. 7507 was sold to the Republic of Korea under the Peace Pheasant program. (David F. Brown)

F-4D 66-7773 marked as the 465 TFS Flagship is seen here at London, Ontario Air Show. It has been "zapped" with the Canadian National Insignia overlaid on the US National Insignia. (David F. Brown)

F-4D MiG killer 66-7550 seen here at Tinker AFB in August 1983 was credited with the destruction of a North Vietnamese MiG-21 shot down on April 15, 1972. (James Goodall)

F-4D MiG killer 66-7550 is seen here at Tinker AFB in September 1988. 7550 is on display at VFW Post 8437 in Enon, Ohio. (Don Spering/AIR)

F-4D MiG killer 66-8714 was credited with the destruction of a North Vietnamese MiG-21 shot down on February 5, 1968. 8714 was transferred to the 301st TFW at Carswell AFB, Texas. (Don Spering/AIR)

MiG Killer F-4D 66-7750 is seen here on the ramp at Tinker AFB on August 17, 1984. 7750 was credited with the destruction of a North Vietnamese MiG-21 shot down on October 24, 1967. (Pete Wilson)

F-4D 66-7608 is seen here at Tinker AFB on May 21, 1981 in the standard Southeast Asia Camouflage scheme. 7608 was sold to the Republic of Korea under the Peace Pheasant program. (Robert J. Pickett)

F-4D 66-7507 is seen here at Tinker AFB on August 17, 1984 in the Southeast Asia Wraparound Camouflage scheme (Pete Wilson)

F-4D 66-7618 is seen here at Tinker AFB on August 17, 1984 in the Southeast Asia Wraparound Camouflage scheme. 7618 ended its career as a Battle Damage Repair aircraft at Tinker AFB. (Pete Wilson)

F-4D 66-7515 is seen here in July 1985 in the Europe I Camouflage scheme. Two maintenance personnel work next to the folded wing. All USAF Phantom IIs, like the Navy versions, had folding wings. (David F. Brown Collection)

F-4D 66-7618 is seen here in January 1988 after being painted in the Europe I Camouflage scheme. (Don Logan Collection)

F-4D 66-7760 is seen here in July 1985 in the Europe I Camouflage scheme. (Paul Osborne)

F-4D 66-7762 is seen here in June 1986 in the Hill Grays Camouflage scheme. 7762 was sold to the Republic of Korea under the Peace Pheasant program. (David F. Brown Collection)

F-4D 66-7773, on the ramp next to 66-7762, having been "zapped" with the emblem of the Montreal Canadians Hockey Team, is seen here in June 1986 painted in the Hill Grays Camouflage scheme. (David F, Brown Collection)

F-4D 66-7515, on the ramp at Luke AFB in November 1988 in the Hill Grays Camouflage scheme. (Kevin Patrick)

Nose Gear Door Art (507th Public Affairs Office)

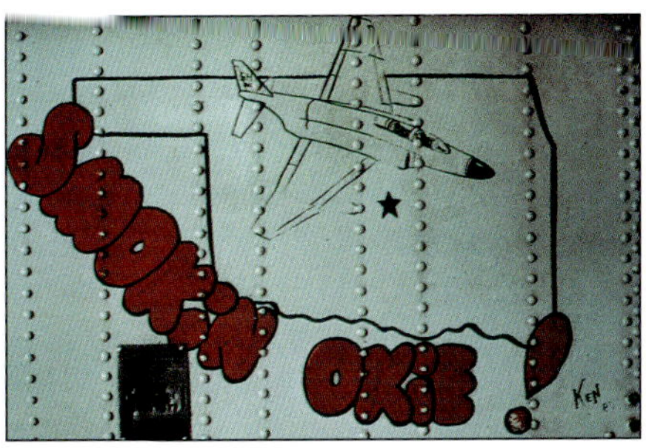

89th TFS 906th TFG, Wright-Patterson AFB, Ohio

Activated on July 1, 1982 as the flying unit of the 906th TFG at Wright-Patterson AFB, Ohio, the 89th TFS flew with **DO** tail codes (for **D**ayton, **O**hio) on their F-4Ds. The F-4Ds had come from the 704th TFS at Bergstrom AFB as the Bergstrom unit converted to LORAN equipped F-4Ds from Kunsan Air Base and Spangdahlem Air Base. Conversion to F-16A/Bs was initiated in October 1989.

906 TFG AFRES

F-4D (Received from the 704 TFS, AFRES, Bergstrom AFB)

66-7506 66-7529 66-7544 66-7550* 66-7554* 66-7582
66-7625 66-7626 66-7650 66-7660 66-7685 66-7687
66-7699 66-7706 66-7718 66-7723 66-7739 66-7749
66-7755 66-7765 66-7771 66-7772 66-7773 66-8825

* 66-7550 - MiG-21 - April 16, 1972
* 66-7554 - MiG-17 - February 14, 1968

F-4D 66-7554 in Southeast Asia Camouflage Scheme, named the "The City of Fairborn" wears two MiG kill stars when photographed at London, Ontario on June 5, 1983. Though 7554 wore two MiG Kill stars for its career with the 89th TFS it was only "officially" credited with the shoot down of a North Vietnamese MiG-17 on February 14, 1968. 7554 was put on display at Wright-Patterson AFB, Ohio. (Paul Osborne)

F-4D 66-7554 in Europe I Camouflage Scheme photographed in August 1984. (James Goodall)

F-4D 66-7554 "The City of Fairborn" in Hill Grays paint scheme in July 1987. (David F. Brown)

F-4D 66-7765 was painted as the 30th Anniversary aircraft representing the 906th TFG at air shows during 1988. (David F. Brown Collection)

F-4D 66-7765 was painted as the 30th Anniversary aircraft representing the 906th TFG at the London, Ontario Air Shows on June 3, 1988. The air show name and a Canadian Maple leaf have been added to the nose for this show. (David F. Brown)

The 30th Anniversary aircraft, 66-7765 represented the 906th TFG at the Dayton, Ohio Air Shows on July 22, 1988. 7765 ended its career as a target on the Avon Park Gunnery Range in Florida. (David F. Brown Collection)

F-4D 66-0259 is seen here in June 1983 in standard Southeast Asia Camouflage. 0259 was put on display at McGuire AFB, New Jersey. (Don Logan Collection)

Above and right: F-4D 66-7582 received a special tail code of TanZ. TanZ was a representation of the name of the 906th TFG's commander Col. David Tanzi. (Pete Wilson)

Below: F-4D 66-8739 is seen here in June 1983 in Southeast Asia Wraparound Camouflage. The DO tail code stands for Dayton, Ohio. (Don Logan Collection)

F-4D 66-7506 is seen here on October 13, 1985 in Europe 1 Camouflage. (J.E. Michaels)

F-4D 66-7650 also in Europe 1 Camouflage is seen here in July 1981. (J.E. Michaels)

F-4D 66-7650 in Hill Grays Camouflage is seen here in September 1986 with the non-standard serial number presentation of 67650. (Don Logan Collection)

F-4D 66-7687 also in Hill Grays Camouflage is seen here at McGuire AFB on October 18, 1987. (Barry Roop)

F-4D 66-7699 "City of Columbus, Ohio" also in Hill Grays Camouflage is seen here on July 18, 1987. (David F. Brown)

704th TFS, 924th TFG, Bergstrom AFB, Texas

Based at Bergstrom AFB, Texas and coming under the 924th TFG, the 704th Tactical Airlift Squadron was redesignated the 704th TFS on July 1, 1981 after completing its conversion from C-130Bs to **TX** (**T**e**X**as) tail coded F-4Ds.

They received their first group of jets in 1981 from the 474th TFW at Nellis AFB followed by additional jets in 1982 from Kunsan AB, Korea and Spangdahlem AB, Germany most of which were equipped with ARN-92 LORAN, distinguishable by the towel rack antenna on the aircraft spine.

The 704th TFS upgraded to F-4Es with ARN-101 DMAS (Digital Modular Avionics System) in 1988. A number of the 704th's F-4Ds were transferred to the Republic of Korea Air Force. In July 1991, the 704th TFS re-equipped with F-16A/Bs.

924 TFG AFRES

F-4D (First group – Received from the 474 TFW, Nellis AFB)

66-7506	66-7529	66-7554*	66-7582	66-7660	66-7685
66-7687	66-7699	66-7706	66-7718	66-7723	66-7739
66-7749	66-7755	66-7765	66-7772		

* 66-7554 - MiG-17 - February 14, 1968

F-4D (Second group – Received from Kunsan/Spangdahlem ABs)

66-7570	66-7641	66-7729	66-8700	66-8709	66-8710
66-8722	66-8732	66-8733	66-8739	66-8740	66-8745
66-8748	66-8762	66-8768	66-8788	66-8797	66-8798
66-8802	66-8805	66-8806	66-8813	66-8819	66-8825

F-4E (Received from the 4 TFW, Seymour Johnson AFB)

72-0135	72-0142	72-1479	72-1483	72-1490	73-1164
73-1166	73-1167	73-1171	73-1184	74-0646	74-0665
74-0666	74-1040	74-1043	74-1621	74-1627	74-1629
74-1630	74-1643	74-1649			

Special Camouflage Scheme

Both F-4D and F-4E aircraft of the 704th TFS carried a special two gray camouflage scheme. This scheme was not applied to any other unit's aircraft. The two grays were Gunship Gray (FS 26118) and Neutral Gray (FS 26270)

(Bob Shane)

(Alec Fushi)

(Don Logan Collection)

F-4D 66-8797 is seen here at Bergstrom AFB on May 22, 1984. It is marked with two command stripes indicating it is the 924th TFG Commanders aircraft. 8797 was sold to the Republic of Korea under the Peace Pheasant program. (Pete Wilson).

F-4E 74-1040 rolls out after touchdown. 1040, painted with the special camouflage used by the 924th TFG, is marked as the 924th TFG Commander's aircraft. (David F. Brown)

MiG Killer F-4D 66-7554 is seen here in April 1982 with two stripes indicating the aircraft is the 924th TFG Commander's aircraft. Though it carries two red stars, it was officially only credited with one MiG kill, a North Vietnamese MiG-17 shot down on February 14, 1968. (Marty Isham Collection)

F-4D 66-7739 is seen here with the dark blue & white checkerboard tail stripe indicating the 704th TFS, 924th TFG is assigned to the 482nd TFW at Homestead. This tail stripe was later changed to a modified version of the Texas State flag. TX was derived from TeXas (Don Logan Collection)

F-4D 66-7685 is seen here on the ramp at Davis-Monthan AFB also marked with the dark blue & white checkerboard tail stripe. (Brian C. Rogers)

F-4D 66-8739 is seen here on April 23, 1983 in the standard Southeast Asia Camouflage Scheme and the Texas flag tail stripe. As indicated by the towel rack antenna, 8739 is equipped with the LORAN navigation system. (Scott Wilson)

F-4D 66-8813 is seen here in the standard Southeast Asia Camouflage Scheme and the Texas flag tail stripe. Many of the 704th TFS F-4Ds were equipped with the LORAN navigation system. (Scott Wilson)

F-4D 66-8739 is seen here in July 1985 painted in Europe 1 Asia Camouflage Scheme. (Paul Osborne)

F-4D 66-8732 is seen here on July 16, 1986 painted in Europe I Asia Camouflage Scheme. (Paul Osborne)

F-4D 66-7641 is seen here in November 1986 at Bergstrom AFB during RAM 86. 7641 is painted in Hill Grays Camouflage Scheme. Visible in the left forward Sparrow well is an AN/AVQ-23A/B Pave Spike LASER designator pod. (George Cockle)

F-4D 66-8813 is seen here on July 16, 1986 at Bergstrom AFB. 8813 is painted in Hill Grays Camouflage Scheme and LORAN equipped. (Brian C. Rogers)

F-4D 66-8813 is seen here on July 16, 1986 at Bergstrom AFB. 8813 is painted in a modified Hill Grays Camouflage Scheme. In the modified scheme the dark gray color continues past the intakes and cockpit almost all the way to the radome. (Brian C. Rogers)

The 704th TFS experimented with a special camouflage scheme of two grays on both its F-4D and F-4E aircraft. F-4D 66-8739 is seen here at Nellis AFB on October 15, 1987 painted in the special scheme. (Pete Wilson).

F-4D 66-8797 is seen at touchdown in September 1988 painted in the special scheme. (David F. Brown).

F-4E 74-1649 is seen in the flare just before touchdown in August 1990. (David F. Brown).

F-4E 73-1164 is seen on the ramp on June 29, 1989. (Douglas E. Slowiak/Vortex Photo Graphics).

F-4E 73-1171 is seen on the ramp in August 1990. (David F. Brown).

F-4E 74-1043 painted in the 704th TFS's special scheme is seen in the flare just before touchdown in August 1990. (David F. Brown).

F-4E 74-1630 painted in the 704th TFS's special scheme is seen in the flare just before touchdown in August 1990. (David F. Brown).

F-4E 72-0135 is seen on the ramp in August 1990. (David F. Brown).

457th TFS, 301st TFW, Carswell AFB, Texas

The 457th was activated as a Tactical Fighter Squadron on July 8, 1972 with F-105D/Fs, replacing the 77th Military Airlift Squadron and its C-124Cs. As part of the 301st TFW, also at Carswell AFB, Texas, the 457th TFS was initially equipped with the Thunderstick II variant of the F-105D, simply called T-Stick IIs. The profile of the T-Stick II caused by the enlarged dorsal fairing containing the special avionics of its aircraft resulted in the squadron being nicknamed The **T**exas **H**umpbacks and assigned the tail code **TH**.

The 457th TFS converted to F-4Ds in 1981 with the jets coming from Kunsan AB, Korea and Spangdahlem AB, Germany. In 1987, the 457th replaced the F-4Ds with F-4Es from the 347th TFW, Moody AFB, Georgia. The 457th TFS in turn retired the Phantoms, replacing them with F-16A/Bs during spring 1991. With the assignment of the F-16s, the tail code was also changed to TF standing for Texas Falcons.

301 TFW AFRES

F-4D (Received from Kunsan/Spangdahlem ABs)

66-7457	66-7511	66-7690	66-7747	66-8709	66-8711
66-8714★	66-8719★	66-8722	66-8727	66-8729	66-8734★
66-8735	66-8737	66-8755	66-8756★	66-8758	66-8759
66-8765	66-8782	66-8786	66-8789	66-8794	66-8803#
66-8806	66-8810	66-8812	66-8816	66-8825	

★ 66-8714 - MiG-21 - February 5, 1968
★ 66-8719 - MiG-17 - December 17, 1967
★ 66-8734 - MiG-21 - May 10, 1972
★ 66-8756 - MiG-19 - May 12, 1972
66-8803 Destroyed in a mishap on May 22, 1987

F-4E (Received from the 347 TFW, Moody AFB)

67-0349	67-0356	67-0364	67-0375	67-0390	67-0392★★
67-0396	68-0308	68-0340	68-0342	68-0347	68-0366
68-0389	68-0391	68-0427	68-0446	68-0447	68-0449
68-0450	68-0452	68-0461	68-0482	68-0497	68-0511
68-0529	68-0532	68-0536			

★★ 67-0392 - MiG-19 - September 2, 1972 - MiG-19 - October 6, 1972

The lead F-4D in standard Southeast Asia Camouflage TFS pitches left with ship two in Southeast Asia Wraparound Camouflage and ship three in Europe I Camouflage ready to follow. (USAF)

F-4D 66-7457, temporarily assigned to the 457th TFS is seen here at Nellis AFB on April 17, 1981 marked as the 457th TFS Flagship. 7457 ended its career as a target on the Holloman AFB target range. (Don Logan Collection)

F-4D 66-8816, ready to release its drag chute, taxis off the runway at Carswell AFB following a local area mission on February 20, 1988. (Brian C. Rogers)

F-4D 66-8816, painted with the 301st TFW command stripe on the fuselage sits on the ramp at Carswell AFB on February 13, 1988. The TH tail code came from the nickname Texas Humpbacks, given to the units F-105D Thunderstick II fighter-bombers assigned to the 301st prior to F-4s. (Keith Snyder)

F-4D 66-8714, on the ramp at Carswell AFB on December 9, 1987 was credited with the destruction of a MiG-21, the shootdown occurring on February 5, 1968. 8714 retired at Carswell AFB, and was put on display. (Brian C. Rogers)

F-4D 66-8794, on the ramp at Carswell AFB on February 20, 1988, through marked with a MiG Kill star on the intake splitter plate, was never officially credited with a shootdown. (Brian C. Rogers)

F-4E 67-0392, painted with the 301st TFW, carried the command stripe on the fuselage. While assigned to the 388th TFW at Korat Air Base in Thailand, 0392 was responsible for the downing of two North Vietnamese MiG-19s, one on September 2, 1972 and the second (shared with F-4E 66-0313) on October 6 of the same year. (James Goodall)

F-4E 67-0392 taxiing out for takeoff in July 1989. 0392 is now on display at the Virginia Air & Space Museum in Hampton, Virginia. (David F. Brown)

F-4E 68-0389, seen here at Carswell AFB on March 9, 1991, is marked for the 301st TFW Fort Worth Phantom Pharewell. (Pete Wilson)

F-4D 66-8714 is seen here on April 30, 1983 in the Southeast Asia Wraparound Camouflage Scheme. As indicated by the towel rack antenna, 8714 is equipped with the LORAN navigation system. Though no MiG kill star was painted on the aircraft when this picture was taken, 8714 is credited with downing a North Vietnamese MiG-21 on February 5, 1968. (Scott Wilson)

F-4D 66-8709 is seen here on April 30, 1983 in the Southeast Asia Wraparound Camouflage Scheme. 8709 is also equipped with the LORAN navigation system. 8709 was used as a target on the Avon Park Gunnery Range in Florida. (Scott Wilson)

F-4D 66-8727 is seen here in April 30, 1983 at Carswell AFB standard Southeast Asia Camouflage Scheme. Visible in the left forward Sparrow well is an AN/AVQ-23A/B Pave Spike LASER designator pod. (Scott Wilson)

F-4D 66-7690 in the Europe I Camouflage scheme is seen in flight on October 22, 1986 carrying an AIM-9 Sidewinder under its left wing. (Brian C. Rogers)

F-4D 66-8735 is seen here on April 30, 1983 in the Europe I Camouflage Scheme, as indicated by the towel rack antenna, 8735 is also equipped with the LORAN navigation system. (Scott Wilson)

F-4D 66-8782 in the Europe I Camouflage scheme is seen at Carswell AFB taxiing out for takeoff on June 5, 1987. (Brian C. Rogers)

F-4D 66-8816 is seen here on December 1, 1984 in the Europe 1 Camouflage Scheme. As indicated by the towel rack antenna, 8816 is also equipped with the LORAN navigation system. (Brian C. Rogers)

LORAN equipped F-4D 66-8816 in Hill Grays Camouflage is seen here in March 1986 with the non-standard serial number presentation of 68816. (Don Logan Collection)

LORAN equipped F-4D 66-8816 in Hill Grays Camouflage is seen here on the ramp at Carswell AFB on August 5, 1986. (Brian C. Rogers)

F-4E 67-0349 in the Europe I Camouflage scheme is seen at Carswell AFB on December 13, 1988, waiting its turn to take the runway for takeoff. 0349 was converted to QF-4E AF-119. (Brian C. Rogers)

F-4E 68-0536 in the Hill Camouflage scheme is seen at Carswell AFB in July, 1990. (Don Logan Collection)

F-4E 68-0536 in the Hill Camouflage scheme is seen at Carswell AFB on January 21, 1989. A QRC 80-01(V) ECM Pod in the forward Sparrow well and two TGM-65 training Mavericks under the left wing. (Keith Snyder)

Air National Guard Units

Thirty-one Air National Guard Squadrons in twenty-three States and the District of Columbia flew versions of the F-4 Phantom II.

SQUADRON	F-4C	F-4D	F-4E	F-4G	RF-4C
106th TRS, 117th TRW, AL ANG					71-94
110th TFS, 131st TFW, MO ANG	79-85		85-91		
111th FIS, 147th FIG, TX ANG	82-86	86-89			
113th TFS, 181st TFG, IN ANG	79-88		88-91		
114th TFTS, 142nd FIG, OR ANG		84-89			
121st TFS, 113th TFW, DC ANG		81-89			
122nd TFS, 159th TFG, LA ANG	79-86				
123rd FIS, 142nd FIG, OR ANG	81-89				
127th TFS, 184th TFG, KS ANG		79-88			
128th TFS, 116th TFW, GA ANG		83-86			
134th TFS, 158th TFG, VT ANG		80-86			
136th FIS, 107th FIG, NY ANG	82-87	87-90			
141st TFS, 108th FW, NJ ANG		81-85	85-91		
153rd TRS, 186th TRG, MS ANG					75-91
160th TRS, 187th TRG, AL ANG					71-83
160th TFS, 187th TFG, AL ANG		82-88			
163rd TFS, 122nd TFW, IN ANG	79-86		86-91		
165th TRS, 123rd TRW, KY ANG					76-89
170th TFS, 183rd TFG, IL ANG	72-81	81-89			
171st FIS, 191st FIG, MI ANG	78-87	87-88			
173rd TRS, 155th TRG, NE ANG					72-94
177th TFTS, 184th TFG, KS ANG		84-90			
178th FIS, 119th FIG, ND ANG	77-90				
179th TRS, 148th TRG, MN ANG					79-83
179th FIS, 148th FIG, MN ANG	83-90				
182ndTFS, 149th TFG, TX ANG		79-86			
184th TFS, 188th TFG, AR ANG	76-88				
189th TRTF, 124th TRG, ID ANG				84-91	
189th FF, 124th FG, ID ANG				91-96	
190th TRS, 124th TRG, ID ANG				75-91	

SQUADRON	F-4C	F-4D	F-4E	F-4G	RF-4C
190th FS, 124th FG, ID ANG				91-96	
192nd TRS, 152nd TRG, NV ANG					75-95
194th FIS, 144th FIW, CA ANG		84-89			
196th TFS, 163rd TFG, CA ANG	82-87		87-90		
196th TRS, 163rd TRG, CA ANG					90-93
199th TFS, 154th CG, HI ANG	76-87				

106th TRS, 117th TRW, Birmingham Municipal Airport, Alabama

The 106th Tactical Reconnaissance Squadron, 117th Tactical Reconnaissance Wing, Alabama Air National Guard was based at Sumpter Smith ANGB, Birmingham MAP. The 106th TRS became the first ANG unit to be assigned Phantom IIs when it received RF-4Cs on February 25, 1971. The unit was equipped with RF-4Cs for over 23 years. The 106th was originally organized in the Alabama National Guard in January 1922 as the 135th Squadron and was given its current numerical designation of 106 in 1924. During World War II, it flew ASW missions in the United States and bombing missions (as the 100th Bombardment Squadron) in the Southwest Pacific Theater. Reorganized postwar as the 106th Bombardment Squadron, the unit became the 106th TRS in February 1951. Prior to receiving its RF-4Cs it had been equipped with RB-26Cs and RF-84Fs. The 106th deployed its RF-84 Thunderflashes to Chaumont AB, France, during the Berlin Crisis.

While equipped with RF-4Cs, the 106th TRS first deployed overseas in March 1976 when it sent aircraft, crews and support personnel to Ramstein AB, Germany. It deployed to the United Kingdom in June/July 1983 (Coronet Joust) and September 1986 (Coronet Mobile). The unit subsequently adopted the BH tail code. The 117th TRW flew at least one F-4C (63-7468).

By early 1990 the 106th TRS was also flying RF-4Cs modified with the Boeing AN/ASQ-203(V)1 Navigation and Weapon Delivery System (NWDS). At the start of Operation Desert Shield in August 1990, three years after becoming the first ANG unit to be equipped with the KS-127A LOROP (Long-Range Oblique Photography) system, the 106th TRS deployed six LOROP-equipped aircraft and personnel to the United Arab Emirates on a voluntary basis. Later relocating to Bahrain, the Alabama volunteers were relieved in the Gulf in December 1990 but left their aircraft (minus 64-1044 lost with its crew in a crash on October 8, 1990) in the theater to be flown and maintained by Nevada Guardsmen during Operation Desert Storm. Another 117th RF-4C, 64-1056 was lost on March 31, 1991 while operating from Bahrain. The loss was due to a bleed air duct failure. Both 192nd TRS Nevada ANG crewmembers ejected safely.

On March 16, 1992, the Birmingham units dropped Tactical from their designation to become the 106th RS, 117th RW. The 117th Reconnaissance Wing received KC-135Rs and was redesignated as the 117th Air Refueling Wing on October 1, 1994.

117 TRW/RW (AL ANG)

RF-4C

63-7745	64-0997	64-1010[*]	64-1032[#]	64-1033	64-1034[*#]
64-1036[#]	64-1038	64-1041[*]	64-1044[#]	64-1046	64-1047[*]
64-1049[*]	64-1051[*]	64-1053[*]	64-1055[*]	64-1056[*]	64-1057[*]
64-1058[*]	64-1060[*]	64-1065[*]	64-1077[*]	64-1082	65-0833[*]
65-0843[*]	65-0854[*]	65-0859	65-0866	64-0867[*]	65-0884[*]
65-0886	65-0893	65-0900[*#]	65-0923[*]	65-0927[*]	65-0945[*]
66-0400	68-0401[*]	66-0415[*]	66-0417	68-0453[*]	68-0575[*]

[*] RF-4C Modified to add AN/ASQ-203(V)1 Navigation and Weapon Delivery System
[#] 64-1032 Destroyed in a mishap on June 30, 1985
[#] 64-1034 Destroyed in a mishap on February 17, 1993
[#] 64-1036 Destroyed in a mishap on May 21, 1985
[#] 64-1044 Destroyed in a mishap on October 8, 1990
[#] 65-0990 Destroyed in a mishap on March 6, 1994

RF-4C 64-1041, painted in the paint scheme for the Phantom's 35th Anniversary, flies with the 117th TRW Commander's aircraft and the 106th TRS Commander's aircraft over the St. Louis Gateway Arch. The BH tail code used by the 117th jets is derived from BirmingHam. (Don Spering/AIR)

RF-4C 65-0854, painted in the paint as the 117th TRW Commander's aircraft is seen here in a climb. (Don Spering/AIR)

A four ship of 117th TRW RF-4Cs, including with the 117th TRW Commander's aircraft and the 106th TRS Commander's aircraft fly over the battleship USS Alabama at it's mooring in Mobile Bay, Alabama. (Don Spering/AIR)

RF-4C 64-0843 is seen here painted in the markings of the 106th TRS Commander's aircraft. (Don Spering/AIR)

A line up of 117th TRW RF-4Cs is seen here on May 4, 1985. (Geoff Lebaron)

RF-4C 65-0854, the 117th TRW Commander's aircraft is seen here marked for the 117th's 20th Anniversary of RF-4C operations. (Don Spering/AIR)

RF-4C 65-0854, the 117th TRW Commander's aircraft, is seen here over Alabama. (Don Spering/AIR)

RF-4C 65-0854 is seen here on the ramp at Lambert Field, St Louis, Missouri at PhanCon 91. 0854 is painted as the 117th TRW Flagship. (Nate Leong)

117th TRW Commanders Flagship RF-4C 65-0854 is seen here on the ramp at George AFB, California at PhanCon 92. 1971 BIRMINGHAM RECCE REBELS 1991 is painted on the upper fuselage. (Don Logan)

RF-4C 64-0843 is seen here painted as the 106th TRS Commander's aircraft. (Don Spering/AIR)

RF-4C 64-0843 the 106th TRS Commander's aircraft is seen here in a dive. (Don Spering/AIR)

RF-4C 65-0843 is seen here on April 24, 1992. 0843 is painted as the 106th TRS Flagship and carries a sharkmouth. (David F. Brown)

RF-4C 65-0833, DAILEY'S RECCE REBELS, is seen here painted as the 106th RS Commander's aircraft. (Don Spering/AIR)

Above: RF-4C 65-0833 is seen here in June 1993 painted as the 106th RS Flagship. It also carries a sharkmouth. 0833 was put on display at VFW Post 4850 in Jasper, Alabama. (David F. Brown)

Right: RF-4C 65-0843 is seen here on August 11, 1992 painted for the Diamond Jubilee (75th Anniversary) of the 106th RS. (Ray Leader)

Below: RF-4C 65-0843 is seen here painted for the Diamond Jubilee (75th Anniversary) of the 106th RS. (Don Spering/AIR)

(Don Spering/AIR)

Markings of the 106th RS 75th Anniversary RF-4C. (Don Spering/AIR)

RF-4C 64-1041 is seen here in flight painted for the 35th Anniversary of F-4 service with the 117th TRW. (Don Spering/AIR)

RF-4C 64-1041 is seen here in June 1993 painted for the 35th Anniversary of F-4 service with the 117th TRW. (David F. Brown)

Markings of RF-4C 64-1041 for the 117th RW 35th Anniversary of operations in the RF-4C. (Don Spering/AIR)

Above and top page 62: RF-4C 64-1057 is seen here marked for the 117th RW Phantom Pharewell in October 1994. (Don Spering/AIR)

Above: RF-4C 64-1057 is seen here marked for the 117th RW Phantom Pharewell in October 1994. (Don Spering/AIR)

Right: Markings of RF-4C 64-1041 for the 117th RW 35th Anniversary of operations in the RF-4C. (Don Spering/AIR)

F-4C 63-7468 was used by the 177th TRW as a pilot training aircraft. (Don Spering/AIR)

F-4C 63-7468 seen here on April 3, 1981 was the only F-4C assigned to the 177th TRW. (Ray Leader)

RF-4C 64-1044 is seen here in February 1979 at Davis-Monthan AFB still carrying the Air National Guard Bi-Centennial emblem on the nose. (Bruce Stewart)

RF-4C 64-1047 is seen here in December 1975 in fresh standard Southeast Asia Camouflage paint scheme. 1047 is now in the USAF Museum collection at Wright-Patterson AFB, Ohio. (D.J. Fisher)

Above and right: 63-7745 the 117th's oldest RF-4C, and RF-4Cs 64-1036 and 64-1047 are seen here over Alabama. (Don Spering/AIR)

RF-4C 64-1053 is seen here on May 5, 1979 in standard Southeast Asia Camouflage paint scheme and a red tail cap. (Jack Morris)

RF-4C 65-0843 is seen here in June 1975 in standard Southeast Asia Camouflage paint scheme. (Don Logan Collection)

RF-4C 65-0893 is seen here on July 18, 1981 in Southeast Asia Wraparound Camouflage paint. (Kirk Minert)

RF-4C 64-1077 is seen here in June 1986 at Luke AFB, Arizona in Europe 1 Camouflage. (Kevin Patrick)

RF-4C 64-1053 is seen here on March 29, 1987 at McGuire AFB, New Jersey in Europe 1 Camouflage. 1053 was converted to a QRF-4C (AF-117) and was shot down on May, 21, 1998. (Barry Roop)

RF-4C 64-1038 is seen here in September 1985. (Don Logan Collection)

RF-4C 64 1047 is seen here at Birmingham on January 31, 1994 with Desert Storm Mission Markings and nose art. The nose art is named IRAQI SCUD SEEKER with additional text which reads "Top Sortie Flyer Desert Shield/Storm 172 Sorties". (Nate Leong)

RF-4C 64-1034 is seen here with the distinctive recce sharkmouth. (Don Spering/AIR)

RF-4C 64-1034 is seen here on October 19, 1991 waiting to take off. 1034 was written off on May 12, 1985 while assigned to the 117th TRW. (Brian C. Rogers)

RF-4C 64-1044 is seen here on May 30, 1987 in the Hill Grays Camouflage Scheme. On October 8, 1990, 1044 became a Desert Shield combat loss. (Donald S. McGarry)

RF-4C 65-0867 ANG PHIRST IN PHANTOMS is seen here over Alabama. (Don Spering/AIR)

RF-4C 65-0867 ANG PHIRST IN PHANTOMS is seen here on the ramp at Birmingham, Alabama. (Don Spering/AIR)

RF-4C 65-0893 is seen here in August 1985 with its rudder carrying the 117th RAM 88 markings. (David F. Brown)

RF-4C 64-1055 is seen here on October 22, 1993 with nose art named BETTER THAN THE BEST – 117th CAMS. (Alec Fushi)

64-1010 DIXIE DEFENDER - Alabama Flag, October 22, 1993 (Pete Wilson)

64-1010 DIXIE DEFENDER - Rebel Flag (Don Spering/AIR)

64-1010 WAR EAGLE Auburn University (Don Spering/AIR)

64-1031 YOU'RE OUT OF PHANTOMS, YOU'RE OUT OF ACTION October 22, 1993. (Don Spering/AIR)

64-1047 IRAQI SCUD SEEKER October 22, 1993. (Pete Wilson)

64-1047 IRAQI SCUD SEEKER Full Nose view. (Don Spering/AIR)

64-1051 CRIMSON TIDE 1992 NATIONAL CHAMPS. (Bill Tate)

Right: 64-1055 BETTER THAN THE BEST – 117th CAMS October 22, 1993. (Pete Wilson)

64-1057 057 FIRST IN LAST OUT. (Don Spering/AIR)

64-1057 SPIRIT OF ALABAMA. (Don Spering/AIR)

64-1058 HUGGIE Teddy Bear. (Don Spering/AIR)

64-1060 HAY MAKER . (Don Spering/AIR)

64-1077 Granny October 22, 1993. (Pete Wilson)

65-0833 DAILEY'S RECCE REBELS. (Don Spering/AIR)

65-0859 BETTY BOOP October 22, 1993. (Pete Wilson)

65-0859 BETTY BOOP and THE JEEP SLAYER Full view. (Don Spering/AIR)

65-0867 ANG PHIRST IN PHANTOMS October 22, 1993. (Pete Wilson)

65-0867 ANG PHIRST IN PHANTOMS Full Nose view. (Don Spering/AIR)

65-0867 UA - TAZ II. (Don Spering/AIR)

65-0884 MYSTIC MISSION MS EMILY Close up. (Don Spering/AIR)

64-0900 UA TAZ October 22, 1993. (Pete Wilson)

64-0900 UA TAZ. (Don Spering/AIR)

64-0923 CITY OF BIRMINGHAM SON O BUCKY. (Don Spering/AIR)

65-0945 #1 Ball October 22, 1993. (Pete Wilson)

65-0945 LUCKY VICKI and #1 Ball Full Nose view. (Don Spering/AIR)

66-0400 KODAK MOMENT. (Don Spering/AIR)

66-0401 HARLEY – DAVIDSON MOTOR CYCLES. (Don Spering/AIR)

66-0415 SWEET THANG II October 22, 1993. (Pete Wilson)

66-0453 WHEN YOU'RE HOT YOU'RE HOT PHANCON–93 SMOKE TRAILS SPECIAL October 22, 1993. (Pete Wilson)

160th TRS/160th TFS, 187th TRG/187th TFG, Dannelly Field, Montgomery, Alabama

From its base at Dannelly Field ANGS at Montgomery Municipal Airport, the 160th, the second Alabama F-4 ANG unit, has flown Phantoms both in the reconnaissance and tactical fighter roles. Designated a fighter squadron and equipped with F-51Ds when extended federal recognition in October 1947. This Alabama unit was redesignated the 160th TRS in September 1950. In the tactical reconnaissance role, the 160th briefly flew RF-51Ds. but was re-equipped with RF-80As shortly after being activated during the Korean call-up. It deployed to Europe in January 1952 and flew RF-80As in Germany and France before being returned to state control and temporarily switching back to RF-51Ds. Thereafter, the 160th TRS successively flew RF-80As (from June 1955) and RF-84Fs (from May 1956).

The 160th TRS received RF-4Cs during the first half of 1971 and first deployed abroad in May/June 1980 when it went to Germany to take part in the NATO BEST FOCUS tactical reconnaissance competition. After flying RF-4Cs for 12 years, the unit received F-4Ds from the 401st TFW, Torrejon AB, Spain and was redesignated the 160th TFS, 187th TFG on July 1, 1983. AL tail codes were applied to 187th TFG F-4Ds. The 160th deployed to Germany in August 1985 for Coronet Meteor, but a little over two years later it began preparations to convert from F-4Ds to F-16A/Bs. The 160th TFS, 187th TFG, officially completed its conversion from F-4Ds to Fighting Falcons on October 1, 1988.

187 TRG/TFG (AL ANG)

RF-4C

63-7741	63-7746	63-7747	63-7750	63-7752	63-7753
63-7754	63-7756[#]	63-7757	63-7758	63-7759	63-7761
63-7762	64-0997	64-1082	65-0822	65-0831	65-0832
65-0833	65-0931	65-0945			

[#] 63-7756 Destroyed in a mishap on February 2, 1983

F-4D (Received from Torrejon AB, Spain)

66-7514 – 01	66-7519 – 02	66-7525 – 03	66-7560 – 04	66-7575 – 05	66-7578 - 06
66-7591 – 07	66-7610 – 08	66-7619 – 09	66-7629 – 10	66-7640 – 11	66-7644 - 12
66-7645 – 13	66-7649 – 14	66-7656 – 15	66-7664 – 16	66-7681	66-7708 - 17
66-7728 – 18	66-7745 – 19	66-7754 – 20	66-8693 - 22		

The 2-digit numbers after the F-4D serial numbers reflect "plane in squadron" numbers that were carried on the nosewheel doors of 160 TFS F-4Ds.

RF-4C 64-0997 is seen here at Dannelly Field in November 1975 in standard Southeast Asia Camouflage. (D. J. Fisher)

RF-4C 64-0997 is seen here at Buckley Air National Guard Base, Colorado on September 1974 in a very weathered standard Southeast Asia Camouflage. (Don Logan Collection)

RF-4C 63-7761 is seen here at Sheppard AFB on February 16, 1974. (Don Logan)

RF-4C 65-0831 is seen here at Dannelly Field on May 6, 1979. (Kirk Minert)

RF-4C 63-7753 is seen here with the USA on the rudder, the markings applied for Best Focus 80. (Ted Van Geffen)

RF-4C 63-7758 also taking part in Best Focus 80. (Ted Van Geffen)

F-4D 66-7578 is seen here in the standard Southeast Asia Camouflage Scheme on June 25, 1983. The AL tail code is derived from ALabama. (Pete Wilson)

F-4D 66-7560 is seen here in the Europe I Camouflage Scheme in August 1984. (David F. Brown)

F-4D 66-7645 is seen here in the Europe I Camouflage Scheme on September 13, 1986. (Brian C. Rogers)

F-4D 66-7649 is seen here in the Hill Grays Camouflage Scheme on June 4, 1988. (David F. Brown Collection)

F-4D 66-7591 is seen here at Dannelly Field in the Hill Grays Camouflage Scheme in December 1986. (Charles T. Robbins)

F-4D 66-7560 is seen here also at Dannelly Field in the Hill Grays Camouflage Scheme in December 1986. (Charles T. Robbins)

184th TFS, 188th TFG, Fort Smith Municipal Airport, Arkansas

Operating from Ebbing Air National Guard Base at the Fort Smith MAP, the 184th TFS flew F-4Cs for 10 years. They deployed to Panama in February 1981 and twice to Turkey (Coronet Crown in September/ October 1983 and Coronet Cherokee in September/ October 1986). Organized after the end of the Korean War, the 184th TRS received federal recognition in October 1953 and was initially equipped with Douglas RB-26Cs. Based at Fort Smith, it converted to jets in 1956, and was equipped with Lockheed RF-80As beginning in June 1956, Republic RF-84Fs beginning in January 1957, and McDonnell RF-101Cs beginning in December 1970.

Reorganized as a fighter unit in 1972, the 184th converted to North American F-100D/Fs during the summer of that year, and it was redesignated 184th TFS in June 1972. F-4Cs replaced F-100s during the summer of 1979 and these Phantoms came from the 56th TFW at Luke AFB, the 35th TFW at George AFB, and the 401st TFW at Torrejon AB, Spain. In April 1989, the 184th TFS completed its conversion to F-16A/Bs.

188 TFG (AR ANG)

F-4C (Received from Luke AFB, George AFB, and Torrejon AB)

63-7411	63-7455	63-7477	63-7516	63-7542	63-7543
63-7682	63-7605	63-7629	63-7637	63-7646	63-7667
63-7672	63-7696#	64-0672	64-0677	64-0725	64-0726
64-0745	64-0748★	64-0761	64-0765	64-0780	64-0784
64-0812	64-0841	64-0912	64-0918	64-0919	

★ 64-0748 - MiG-21 - May 20, 1967
63-7696 Destroyed in a mishap on June 5, 1985

Above and right: F-4C MiG killer 64-0748 is seen here carrying FS (Fort Smith) Tail codes and marked with artwork by Don Spering as the CITY OF FT. SMITH ARKANSAS. 0748 was put on display at Little Rock AFB, Arkansas. (Don Spering/AIR)

Below: F-4Cs of the 188th TFG are photographed on the Nellis AFB ramp on October 15, 1985 during Gunsmoke 85. (Marty J. Isham)

F-4C MiG killer 64-0748, photographed on January 15, 1985 in Southeast Asia Wraparound Camouflage was credited with the destruction of a North Vietnamese MiG-21 shot down on May 20, 1967. (Mark Bacon)

F-4C MiG killer 64-0748, photographed on June 20, 1987 in Hill Grays. (Pete Wilson)

F-4C 63-7411, the 6th F-4C built, was the oldest F-4 in Tactical Air Command Service when photographed in May 1988. (Don Spering/AIR)

F-4C 63-7411 is seen here marked with artwork by Don Spering recognizing it as the oldest F-4 in TAC (Tactical Air Command). 7411 was put on display at Fort Smith, Arkansas. (Don Spering/AIR)

F-4C 63-7455 is seen here at Luke AFB in February 1983 in the standard Southeast Asia Camouflage scheme. (Kevin Patrick)

F-4C 64-0919 also seen here at Luke AFB was photographed in October 1986 in the Southeast Asia Wraparound Camouflage scheme. (Kevin Patrick)

F-4C 64-0919 seen here taxiing at Nellis AFB on October 5, 1983. (Don Logan Collection)

F-4C 63-7605 also seen here at Luke AFB was photographed in October 1986 in the Southeast Asia Wraparound Camouflage scheme. (Kevin Patrick)

F-4C 64-0780 is seen here at Luke AFB with the CORNET PAWNEE NORWAY 86 emblem on the rudder. (Kevin Patrick)

F-4C 64-0672 is seen here at Carswell AFB, photographed on January 11, 1987 with the CORNET CHEROKEE emblem on the rudder. (Brian C. Rogers)

F-4C 64-0768 also seen here at Luke AFB was photographed in October 1986 in the Europe 1 Camouflage scheme. (Douglas E. Slowiak/Vortex Photo Graphics)

F-4C 63-7637 also seen here at Luke AFB was photographed in October 1986 in the Southeast Asia Wraparound Camouflage scheme. (Kevin Patrick)

194th FIS, 144th FIW, Fresno ANGB, California

The 194th FIS was equipped with F-4Ds for six years starting in late 1983. Based at the Fresno ANGB, Fresno Air Terminal. The 194th Fighter Squadron (SE) was federally recognized in March 1949 and was equipped with F-51Ds and F-51Hs until the summer of 1953, when it converted to F-86As. The switch to jets required the relocation of the 194th from Hayward to Fresno in November 1954. The 194th FIS has remained in central California and has not been called to active duty. It has successively converted to F-86Ls in April 1958, to F-102A/TF-102As in July 1964, and to F-106A/Bs during summer 1963.

The 194th FIS converted to F-4Ds during the winter 1983 receiving aircraft from the 31st TFW at Homestead AFB. The unit also maintained an alert detachment of F-4Ds at George AFB, California. Between April 1986 and April 1987, the 194th FIS and its parent, 144th FIW, provided F-4Ds and personnel to Operation Creek Klaxon, the Air National Guard's assumption of air defense commitment at Ramstein AB, Germany, while the 86th TFW, USAFE, was converting to F-16s. The 144th FIW received F-16A/B ADFs in summer 1989. Still retaining air defense as its primary role, the Fresno units were redesignated 194th FS, 144th FW, on March 16, 1992.

144 FIW (CA ANG)

F-4D (Received from Homestead AFB)

65-0583 65-0584 65-0588 65-0596 65-0611 65-0646
65-0662 65-0667 65-0694 65-0712 65-0721 65-0740
65-0747 65-0760 65-0763 65-0775 65-0779 65-0789
66-0262 66-0276 66-0279 66-7487 66-7668 66-7698
66-7741 66-8728 66-8817

F-4C 65-0646 is seen here at Luke AFB on August 24, 1985 in the Southeast Asia Wraparound Camouflage scheme. (Douglas E. Slowiak/Vortex Photo Graphics)

F-4C 65-0646 is seen here at Luke AFB in October 1985 in the Southeast Asia Wraparound Camouflage scheme. The aircraft is carrying 370 gallon external wing tank painted ADC Gray. (Craig Kaston)

F-4C 65-0662 is seen here at Luke AFB in December 1987 in the Europe 1 Camouflage scheme. (Marty J. Isham)

F-4C 65-0712 is seen here in April 1987 in the Europe 1 Camouflage scheme. (David F. Brown)

F-4C 65-0763 is seen here on June 23, 1984 in Air Defense Command Gray paint scheme. Note the unusual USAF and serial number presentation on the vertical fin and the blue stripe on the external wing tank. (Brian C. Rogers)

F-4C 65-0596 is seen here in April 1985 in Air Defense Command Gray paint scheme. (David F. Brown)

F-4C 66-0276 is seen here in May 1986 in Air Defense Command Gray paint scheme. (Don Logan Collection)

F-4C 65-0760 is seen here in July 1988 in Air Defense Command Gray paint scheme with the Griffin of the 144th FIW painted on the vertical fin. (Don Logan Collection)

F-4C 66-7668 is seen here at Fresno in September 1992 with the Griffin of the 144th FIW painted on the vertical fin. (Paul Minert)

F-4C 66-7741 is seen here at Fresno in October 1988 in Hill Grays with the Griffin of the 144th FIW painted on the vertical fin. (David F. Brown)

F-4C 65-0646 is seen here at Fresno on April 17,1988 in Hill Grays with the Griffin of the 144th FIW painted on the vertical fin. (Don Logan Collection)

F-4C 65-0667 is seen here on April 17,1988 in Hill Grays. (James Goodall)

196th TFS/196th TRS, 163rd TFG/163rd TRG, March AFB, California

The 196th Tactical Reconnaissance Squadron, 163rd Tactical Reconnaissance Group, California Air National Guard was based at March Air Force Base, 9 miles southeast of Riverside, California. After flying F-51Ds since its organization in November 1946, the 196th FS became the first Guard unit to convert to jets when it received F-80Cs directly from the Lockheed assembly line in June 1948. Called to active duty in October 1950, the 196th Fighter-Bomber Squadron converted to F-84Es prior to deployment to Japan and subsequent combat operations in Korea. Following its return to Ontario Airport in Southern California in July 1952, the squadron was re-equipped with F-51Hs in October 1952 and redesignated 196th Fighter Interceptor Squadron in January 1953. The 196th FIS converted to F-86As in March 1954, to F-86Ls in February 1961 and to F-102A/TF-102As in May 1965. In March 1975, the 196th retired its Convair F-102A/TF-102As, converted to Cessna O-2As and was redesignated the 196th TASS. After flying the O-2 aircraft for over seven years, the squadron moved from Ontario IAP to March AFB during the summer of 1982.

On October 1, 1982 it was redesignated 196th TFS from 196th TASS on 1 October 1982 and began conversion to F-4Cs received from the 58th TFW at Luke AFB. While flying F-4Cs, the squadron deployed to Spain in May/June 1986 for Coronet Laguna. The unit converted to F-4Es during the spring of 1987. The F-4Es came from

163 TFG/TRG/RG (CA ANG)

F-4C (Received from Luke AFB)

63-7418	63-7426	63-7432	63-7473	63-7485	63-7497
63-7510	63-7516	63-7519	63-7522	63-7545	63-7553
63-7559	63-7569	63-7611	63-7617	63-7644	63-7654
63-7655	63-7665	63-7686	63-7689	63-7693	63-7711
64-0665	64-0677	64-0682	64-0783	64-0789	64-0796
64-0827	64-0889	64-0910	64-0923#		

64-0293 Destroyed in a mishap on March 21, 1987

F-4E (Received from George AFB and Moody AFB)

67-0345	67-0348	67-0353	67-0360	67-0372	67-0384
67-0394#	67-0395	68-0317	68-0318	68-0320	68-0321
68-0324	68-0343	68-0358	68-0363	68-0367#	68-0382
68-0384	68-0387	68-0390	68-0402	68-0404	68-0409
68-0423	68-0424	68-0426	68-0443	68-0466	68-0490
68-0507	68-0512#	68-0514	68-0515	69-0249	69-0291★
69-0307	69-7267				

★ 69-0291 - MiG-21 - August 19, 1972
67-0394 Destroyed in a mishap on July 6, 1989
68-0367 Destroyed in a mishap on July 27, 1989
68-0512 Destroyed in a mishap on June 13, 1990

RF-4C (Received from Bergstrom AFB and Other Bases)

63-7742	63-7746	63-7750	63-7757	63-7762	64-0997*
64-1010*	64-1053	64-1075*	65-0852*	65-0901*	65-0918*
65-0922*	65-0927	65-0942*	66-0393*	66-0397*	66-0401*
66-0402*	66-0407*	66-0417*	66-0422*	66-0423*	66-0435*
66-0444*	66-0452*	66-0453*	66-0461*	66-0475	67-0429
67-0433	67-0434*	67-0444	67-0449*	67-0454	67-0469

* RF-4C Modified to add AN/ASQ-203(V)1 Navigation and Weapon Delivery System

(Paul Minert)

the 35th TFW at George AFB and the 347th TFW at Moody AFB, Georgia. Switching to the reconnaissance role, in the spring of 1990 the 196th converted to RF-4Cs. Many of the RF-4Cs came from Bergstrom AFB. It was redesignated 196th TRS, 163rd TRG on July 1, 1990. Like other ANG RF-4C reconnaissance units the 196th flew aircraft equipped with the Boeing NWDS. Both units dropped the Tactical from their designation becoming the 196th Reconnaissance Squadron of the 163rd Reconnaissance Group on March 16, 1992. The 163rd Reconnaissance Group gave up their RF-4Cs and received KC-135Es and, as a result, was redesignated as the 163rd Air Refueling Wing on October 1, 1993, The Unit transitioned to KC-135Rs during late 1996 and early 1997.

Below: F-4E 67-0372, painted as the 163rd TFG Flagship on the ramp at March AFB in December 1989. (Paul Minert)

F-4E 69-0291 is seen here in June 1988 painted in Europe 1 Camouflage scheme with a MiG kill star on the right splitter plate. While assigned to the 388th TFW at Korat Air Base in Thailand, 0291 was responsible for the downing of a North Vietnamese MiG-21 on August 19, 1972 while assigned to the 366th TFW. 0291 was later converted to an F-4G. (Paul Minert)

F-4C 63-7426 is seen here at March AFB on September 25, 1983 painted in the Southeast Asia Wraparound Camouflage scheme. 7426 was later transferred to Sheppard AFB, Texas where it served as a GF-4C Ground Maintenance Trainer. (Donald S. McGarry)

F-4C 63-7497 is seen here at March AFB on November 13, 1983 with BDU-50 AIR bombs (Inert MK 82 500 pound bomb). BDU is the abbreviation for Bomb Dummy Unit. 7497 is painted in the Southeast Asia Wraparound Camouflage scheme. The aircraft was later used as a target on the Nellis AFB Range Complex. (Donald S. McGarry)

F-4C 64-0682 is seen here at March AFB on September 25, 1983 painted in the Southeast Asia Wraparound Camouflage scheme. (Donald S. McGarry)

F-4C 64-0827 is seen here at March AFB in March 1983 painted in the Southeast Asia Wraparound Camouflage scheme. The CALIFORNIA tail stripe, lacking the edge markings is slightly different than standard. (Scott Wilson)

F-4C 63-7473 is seen here in November 1985 in the Europe I Camouflage scheme. An AIM-9 Training missile with out fins is loaded on the right inboard pylon. The aircraft was later used as a target on the Nellis AFB Range Complex. (Don Logan Collection)

F-4C 63-7644 is seen here at March AFB in August 1986 in the Europe I Camouflage scheme. A travel pod is loaded on the right inboard pylon. The aircraft was put on display at Arnold AFB, Tennessee. (James Goodall)

F-4C 63-7553 in a vertical climb. (Don Spering/AIR)

F-4C 64-0677 is seen here visiting Luke AFB in August 1985 in the Europe I Camouflage scheme. (Kevin Patrick)

F-4C 63-7689 is seen here visiting Luke AFB in December 1986 painted in the Hill Grays Camouflage scheme. 7689 is marked with the five-digit serial number presentation. 7689 was later flown by Tracor Flight Systems Inc. at Mojave, California as N420FS. (Kevin Patrick)

F-4C 64-0682 is seen here at March AFB on April 16, 1986 painted in the Hill Grays Camouflage scheme. 0682 was later used as a battle damage repair aircraft at Seymour-Johnson AFB, North Carolina. (Ray Leader)

F-4C 63-7545 is seen here at March AFB in August 1986 in the Hill Grays Camouflage scheme. The camouflage pattern differs slightly on each individual F-4. On 7545 the dark gray on the nose continues up to the edge of the radome. Starting in July 1987, 7545 was used by Tracor Flight Systems Inc. at Mojave, California flying as N421FS. (James Goodall)

F-4E 68-0320 is seen here at March AFB in April 1992 in a weathered Europe 1 Camouflage scheme. 0320 was converted to QF-4E AF-123. (James Goodall)

F-4E 68-0321 is seen here at March AFB on November 20, 1987 in Europe 1 Camouflage scheme. BDU-33 25-pound practice bombs are visible on the inboard pylons. The BDU-33s, like all inert practice weapons, are painted blue. (Brian C. Rogers)

F-4E 68-0466 is seen here at March AFB in April 1990 in Europe I Camouflage scheme. 0466 was transferred to the Republic of Korea Air Force in September 1990. (Paul Minert)

F-4E 68-0348 is seen here at March AFB in April 1990 in Hill Grays paint scheme. (James Goodall)

F-4E 68-0367 is seen here taxiing on October 9, 1987 in Hill Grays paint with a travel pod and two 370 gallon external wing tanks. 0367 crashed on July 27, 1989 while assigned to the 163rd TFG. (Brian C. Rogers)

F-4E 67-0353 is seen here at March AFB on June 18, 1988 with BDU-33 practice bombs and the high speed center line fuel tank. 0353 was transferred to the Republic of Korea Air Force on February 23, 1989. (Brian C. Rogers)

RF-4C 67-0353 is seen here taxiing back after landing at March AFB in Hill Grays paint. For all full stop landings, it was standard practice for USAF F-4s to use the drag chute, seen here trailing behind the aircraft. (Douglas E. Slowiak/Vortex Photo Graphics)

RF-4C 64-1053 is seen here at March AFB on March 22, 1992 in Hill Grays paint. It is carrying a travel pod and a high-speed centerline tank. 1053 was converted to QRF-4C AF-117 and shot down. (Kevin Patrick)

RF-4C 66-0444 is seen here at March AFB in May 1990 in Hill Grays Camouflage. (Ben Knowles)

RF-4C 66-0435 is seen here at March AFB in November 1991 with the crew on board and engines running, ready to taxi on a training mission. (Paul Minert)

RF-4C 66-0407 is seen here on the ramp at March AFB in April 1992 in Hill Grays paint. (Paul Minert)

121st TFS, 113th TFW, District of Columbia ANG, Andrews AFB, Maryland

Based at Andrews AFB, Maryland, the 121st TFS, 113th TFW, District of Columbia Air National Guard flew its first F-4D sorties in July 1981. Most of the aircraft had come from Kunsan AB, Korea and Kadena AB, Japan. Unlike other Air National Guard units which have a dual state/federal status, with the governors of their state as their commanders-in-chief in peacetime, the District of Columbia ANG comes directly under the Commanding General of the District of Columbia. When placed on federal active duty, the DC ANG like all Guard units have the President of the United States as their commander-in-chief. The flying unit of the District of Columbia was organized on April 10, 1941 as the 121st Observation Squadron and, redesignated 121st Liaison Squadron, took part in combat operations in Algeria, Italy and France during World War II.

Following the end of the World War II, the squadron had been assigned F-47Ds, F-84Cs, F-94Bs (on active duty during the Korean War), F-51Hs, F-86Es, F-86Hs, F-100Cs (with which it was activated during both the 1961-1962 Berlin Crisis and the 1968-1969 Pueblo Crisis), F-105D/Fs, F-4Ds and F-16A/Bs.

While equipped with F-4Ds, the 121st TFS deployed to the United Kingdom in June 1983 (Coronet Shield) and to Iceland in April/May 1986 (Coronet Kiowa). In 1984, it became the first Guard unit to be trained with the Pave Spike system and Laser Guided Bombs. The 121st TFS received its first F-16As in September 1989, and completed its conversion from Phantoms to Fighting Falcons during the fall of 1990.

113 TFW (DC ANG)

F-4D (Received from Kunsan AB and Kadena AB)

66-7509	66-7524	66-7531	66-7538	66-7547	66-7549
66-7556	66-7561	66-7588	66-7607	66-7614	66-7623
66-7634	66-7658	66-7661**	66-7666	66-7667	66-7677
66-7683	66-7684	66-7688	66-7693	66-7702	66-7711
66-7712	66-7722	66-7746	66-8783	66-8821	

** 66-7661 - MiG-17 - February 14, 1968 - MiG-21 - May 11, 1972

F-4D 66-7538 in a climb. The DC tail code comes from Washington DC. (Don Spering/AIR)

F-4D double MiG Killer 66-7661 is seen here over Andrews AFB, Maryland in June 1984. (Frank MacSorley)

F-4D 66-7561 painted in Europe I Camouflage scheme escorts 66-7614 painted in Hill Grays when photographed on May 30, 1987. Aircraft painted in both the Europe I scheme and the Hill Grays I scheme were possessed by the 113th TFW at the same time. (Barry Roop)

F-4Ds 66-7712 and 66-7666 (both in standard Southeast Asia Camouflage scheme) climb out following a formation takeoff from Andrews AFB in May 1982. (David F. Brown)

The tail of F-4D 66-7693 carried 113th TFW Commanders markings when photographed at Andrews AFB in February 1989. In addition, the Commander's markings included the distinctive red vertical stripe on the intake lip. 7693 was retired and put on display at the DC air National Guard, Andrews AFB, Maryland. (David F. Brown)

F-4D 66-7693 carried a red Commander stripe on the Southeast Asia Wraparound Camouflage scheme when photographed at Andrews AFB on May 14, 1983. (Frank MacSorley)

The red star on the right intake splitter of F-4D 66-7661 confirms it was credited with the downing of a North Vietnamese MiG. It was actually credited with two MiGs, a MiG-17 on February 14, 1968 and a MiG-21 on May 11, 1972. An identical star is also carried on the left intake splitter. 7661 was photographed at Andrews AFB on November 22, 1986 in a weathered Europe 1 Camouflage scheme. (Geoff Lebaron)

F-4D 66-7661 is seen here in Hill Grays Camouflage with the MiG kill star when photographed at Andrews AFB in November 1989. 7661 is carrying BDU-33 practice bombs. (Bob Shane)

F-4D 66-7712 is painted in a standard Southeast Asia Camouflage scheme. The light brown paint on the tail is somewhat weathered and had been touched up when photographed at Andrews AFB on March 2, 1985. 7712 is carrying 370 gallon drop tanks painted in the wraparound version of the Southeast Asia Camouflage. (David F. Brown Collection)

F-4D 66-7693 is painted in a Southeast Asia Wraparound Camouflage scheme. Having just arrived at Andrews AFB, the DC tail code of the 113th TFW has been applied, but the tail stripe and other unit markings have not yet been added when photographed in October 1981. A 113th TFW F-105D (the aircraft type assigned to the 113th before F-4s) is visible in the background. (Don Logan Collection)

F-4D 66-7722 is painted in the Southeast Asia Wraparound Camouflage scheme. 7722 is wearing a non-standard version of the 113th TFW tail stripe when photographed in November 1982. (Don Logan Collection)

F-4D 66-7607 is painted in Europe I Camouflage scheme when photographed on June 3, 1988 at the London, Ontario Air Show. It also wears the sculpted black nose paint from the radome and over the canopies. (David F. Brown)

F-4D 66-7538 is painted in Europe I Camouflage scheme when photographed at Andrews AFB in March 1984. As evident by the adjacent F-4, aircraft painted in both the Southeast Asia scheme and the Europe I scheme were possessed by the 113th TFW at the same time. (David F. Brown)

MiG Killer F-4D 66-7661 is seen over a snow covered Maryland countryside in December 1989. (Don Spering/AIR)

F-4D 66-7623 is painted in Hill Grays paint scheme. (Bob Shane)

F-4D 66-7688 is painted in Hill Grays paint scheme when photographed in May 1986. 7688 is wearing the non-standard five digit serial number. Under the right wing are Aim-9 Sidewinders and GBU-12 Paveway II Laser Guided Bombs. An inert AIM-7 Sparrow is visible in the right forward missile well and a high-speed fuel tank is on the centerline station. (David F. Brown)

F-4D 66-7688 is seen here carrying BDU-33 when photographed in October 1986. (Don Logan Collection)

128th TFS, 116th TFW, Dobbins AFB, Georgia

Flying Phantoms for only four years, the 128th TFS had a 50 year history flying from Dobbins ARB in Marietta, Georgia. Organized in May 1941 as the 128th Observation Squadron, the first flying unit in the Georgia National Guard was almost immediately placed on active duty. It continued on active duty, flying ASW missions over the Gulf of Mexico and bombing missions from Italy as the 840th Bombardment Squadron equipped with B-17Gs. Following World War II, it was reorganized as a fighter squadron flying P-47Ns, F-51Hs, F-84Ds, F-84Fs and F-86Ls, before converting to C-97Fs as the 128th Air Transport Squadron. As a transport unit, it flew C-124Cs between 1966 and 1973. In April 1973 it once again became a fighter squadron and was re-equipped with F-100Ds. F-105G Wild Weasels replaced the F-100s in 1979.

The 128th TFS began its conversion from Wild Weasel F-105Gs to F-4Ds in January 1983 and completed the process in less than nine months receiving the aircraft from the 401st TFW, Torrejon AB, Spain. The 128th TFS did manage to deploy overseas in August 1985, when its F4Ds went to Germany (Coronet Meteor). Remaining at Dobbins AFB the 128th officially completed its conversion to F-15A/Bs on March 31, 1987. As part of its conversions to B-1Bs in 1996, the unit was relocated 125 miles south to Robins AFB in Warner Robins, Georgia. On April 1, 1996, the unit was officially redesignated as the 128th Bomb Squadron, 116th Bomb Wing and began its new mission with the B-1B Lancer.

116 TFW (GA ANG)

F-4D (Received from Torrejon AB)

66-7539	66-7553	66-7566	66-7579	66-7588	66-7595
66-7604	66-7605	66-7607	66-7614	66-7623	66-7662
66-7676	66-7689	66-7694	66-7702	66-7711	66-7714
66-7720	66-7721	66-7735	66-7746	66-7759	66-7768
66-8689#	66-8783	66-8808			

66-8689 Destroyed in a mishap on September 24, 1983

Georgia ANG F-4Ds, 66-7662 escorts F-105G Wild Weasel 63-8299 on the Georgia Air National Guard's last flight of a F-105G on May 25, 1983. The conversion to Phantom II from F-105Gs began in January 1983. (Don Spering/AIR)

Two Georgia ANG F-4Ds, 66-7605 and 66-77607, escort F-15A 74-0128, a new addition to the 118th TFW. The conversion to F-15As, replacing the F-4Ds, was completed by the 118th TFW in March 1987. (Don Spering/AIR)

F-4D 66-7553 is seen here on the ramp with a Vermont ANG F-4D on May 14, 1983 wearing the yellow Georgia tail stripe used on their F-105s and early F-4s. 7553 was painted in the Southeast Asia Wraparound Camouflage scheme in these photographs taken at Andrews AFB. (Frank MacSorley)

F-4D 66-7588 was painted in the Southeast Asia Wraparound Camouflage scheme when photographed here on the ramp in September 1983. (Don Logan Collection)

With the crew onboard, F-4D 66-8808 readies to depart the transient line when photographed here on November 10, 1984. 8808 was later used on the Avon Park Gunnery Range, Florida as a target. (Brian C. Rogers)

F-4D 66-7614, in Europe I Camouflage scheme wears the new GEORGIA tail cap markings when photographed here on November 3, 1985. This tail cap marking was later applied to their F-15s and B-1Bs. (Brian C. Rogers)

F-4D 66-7539, in Europe I Camouflage scheme, wears the new GEORGIA tail cap markings when photographed here in April 1986. (James Goodall)

199th TFS, 154th Composite Group, Hickam AFB, Hawaii

Based at Hickam AFB, the Hawaii ANG operated F-4Cs in an Air Defense role assigned to the 199th TFS (nicknamed "The Kukaiumoku") of the 154th FW. The unit was one of the last users of the F-4, replacing them with F-15A/Bs in the summer of 1987.

The 199th was organized in September 1946 as the 199th FS, receiving P-47Ns in July 1947, It was redesignated the 199th FIS in 1952 and converted to F-86Es in 1954. To provide an all- weather air defense, the 199th FIS received F-86Ls in 1958 and F-102As in 1961. The F-102s were replaced by F-4Cs in 1976. In June 1976 the squadron was redesignated the 199th TFS, even though air defense was still its primary mission.

While flying F-4Cs, the 199th TFS made three overseas deployments, going to Japan in June 1980 (Cope North) and twice to the Philippines for Cope Thunder exercises (April/May 1981 and May 1985). The 199th TFS began exchanging its F-4Cs for F-15A/Bs in March 1987 and completed the conversion process in January 1988. The majority of the original group of F-15 aircraft came from the 43rd TFS, 21st TFW at Elmendorf, as that unit transitioned from the F-15A/B to the F-15C/D. The original group of F-15 aircraft was exchanged for F-15A/B MSIP aircraft from Holloman in 1992.

154 COMPG (HI ANG)

F-4C

63-7414	63-7415	63-7454	63-7463	63-7491	63-7505
63-7511	63-7540	63-7552	63-7562	63-7575	63-7578
63-7585	63-7588	63-7592	63-7598	63-7625	63-7628
63-7632	63-7647★	63-7649	63-7676	63-7705	64-0711
64-0715	64-0785	64-0792	64-0793	64-0806	64-0831
64-0851	64-0902	64-0913	64-0914		

★ 63-7647 - MiG-17 - June 5, 1967

F-4C 63-7647, nicknamed ALAE OLA, is seen here on the ramp at Hickam AFB on July 29, 1984. While assigned to the 8th TFW at Ubon RTAFB in Thailand, 7674 was responsible for the downing of a North Vietnamese MiG-17 on June 5, 1967. The second star is for a kill which was not confirmed by the USAF. The aircraft was retired to the Oregon Air & Space Museum. (Robert F. Dorr)

F-4C 64-0793, in a fresh standard Southeast Asia Camouflage scheme, is seen here without unit markings on October 13, 1976. 0793 was used a battle damage repair aircraft at Hickam AFB and later scrapped. (Bill Curry)

F-4C 63-7676, with the distinctive red yellow and black HAWAII tail stripe and a Southeast Asia Wraparound Camouflage scheme is seen here on the ramp at Hickam AFB on December 27, 1984. 7676 was used as a battle damage repair aircraft at Torrejon Air Base, Spain and later scrapped. (Marty Isham Collection)

F-4C 64-0902 is seen here on the ramp at Hickam AFB in December 1984. (David F. Brown)

F-4C 64-0715 is seen here on the ramp at Hickam AFB in December 1984 as the aircrew prepares to start engines. 0715 was used a battle damage repair aircraft at Eglin AFB and later scrapped. (David F. Brown Collection)

F-4C 63-7575, also wearing the distinctive red yellow and black HAWAII tail stripe, is seen here on the ramp at Hickam AFB on May 16, 1987 in Europe 1 Camouflage. 7575 was later used as a range target at Holloman AFB, New Mexico. (Marty Isham Collection)

F-4C 63-7511, carrying the Hawaiian name HONAKAI, painted in Hill Grays scheme is seen here on the ramp at Hickam AFB on May 16, 1987. 7511 was used as a battle damage repair aircraft at Clark Air Base, Philippines and later scrapped. (Marty Isham Collection)

F-4C 64-0806, lit by and early morning sun, is seen here on May 9, 1985. The aircraft is now on display at Nellis AFB, Nevada. (Brian C. Rogers)

F-4C 64-0913, in Hill Grays Camouflage scheme, is seen here on the ramp at Hickam AFB in January 1986. Of note is the non-standard placement on the tail of the five digit serial number. 0913 is now on display at Kadena Air Base, Okinawa, Japan. (Jerry Geer)

With the distinctive red yellow and black HAWAII tail stripe and painted gloss gray, F-4C 63-7414, is seen here taxing on June 17, 1981. 7414 carries the Air Defense style large five digit tail number. 7414 is on display at the Midland Air Museum in England. (James Goodall)

F-4C 63-7562, wearing the Hawaiian name OLILI, is seen here on the ramp at Nellis AFB on March 6, 1982. (Ray Leader)

F-4C 64-0785 in gloss gray is seen here on the ramp at Hickam AFB on July 29, 1984. The Unit insignias are visible on the intake splitter. 0785 was used a battle damage repair aircraft at Hill AFB, Utah. (Robert F. Dorr)

F-4C 64-0792, with the wings folded and a recently painted rudder is seen on the ramp at Hickam AFB on September 1982. 0792 was later placed on display at Hickam AFB. (Don Logan Collection)

F-4C 64-0851 is seen here on the ramp at Hickam AFB on December 1984. 0851 was placed on display at Tuskegee Institute, Tuskegee, Alabama. (Don Logan Collection)

F-4C 63-7540 is seen here on the ramp at Andersen AFB, Guam on April 24, 1981. 7540 carries the Air Defense style large five digit tail number. 7540 was placed on display at Hickam AFB, Hawaii. (David F. Brown)

190th TRS/FS, 124th TRG/FG, Gowen Field, Boise, Idaho

The 190th Fighter Squadron, 124th Fighter Wing, Idaho Air National Guard, located at Gowen Field, Boise Air Terminal, outside Boise, sent their last F-4, (the last combat F-4 in the U.S. military), an F-4G Wild Weasel to AMARC at Davis-Monthan on April 20, 1996. The unit had operated RF-4Cs since the fall of 1975, receiving their first F-4G for conversion in the summer of 1991.

Equipped with F-51Ds, the Idaho unit was federally recognized as the 190th Fighter Squadron (SE) in October 1946. It served on active duty in the United States during 1951 and 1952. It was reorganized as the 190th Fighter Interceptor Squadron in January 1953. and converted to F-86As in November 1953. The 190th again changed aircraft to F-94Bs in December 1954, to F-89Bs in July 1956, to F-86Ls in April 1959 and to F-102As in April 1964.

In October 1975, the unit became the 190th Tactical Reconnaissance Squadron and exchanged its Delta Daggers for RF-4Cs. The 190th TRS shared the RF-4Cs with the Reconnaissance Weapons School, organized in March 1983 to teach advanced reconnaissance tactics to USAF and ANG crews, and the 189th Tactical Reconnaissance Training Flight (TRTF), activated on September 1, 1984 to take over RF-4C training responsibility from the 190th TRS.

The 190th converted from RF-4Cs to F-4Gs during the summer of 1991 and was redesignated the 190th Fighter Squadron on 16 March 16, 1992. The Reconnaissance Weapons School was inactivated on December 31, 1991 and the training unit, training both RF-4C and F-4G crews, was redesignated the 189th Fighter Flight on March 16, 1992.

Idaho ANG began converting to the A-10 when their first aircraft arrived from the Massachusetts Air National Guard on March 20, 1996. The 190th conversion to C-130s for the Airlift Squadron occurred at the same time, making the 124th Wing a composite wing with a squadron of A-10s and one of C-130s.

189th TRTF patch

124 TRG/FG (ID ANG)

RF-4C

63-7742*	63-7754*	64-1031*	64-1035*	64-1060*	64-1061
64-1063*	64-1067	64-1071	64-1074	64-1075*	64-1080*
65-0823*	65-0826	65-0831	65-0837	65-0845	65-0849
65-0852*	65-0869	65-0874*	65-0879*	65-0892*	65-0896
65-0898*	65-0900*	65-0901*	65-0902*	65-0904*#	65-0907*
65-0908*	65-0910*	65-0911*	65-0912*	65-0918*	65-0922*
65-0923*	65-0924	65-0928*	65-0929#	65-0944	66-0399*
66-0425*	66-0430*	66-0446	67-0428	67-0453	67-0463
67-0466*	68-0569	68-0577	68-0594	68-0595	68-0596
68-0599	68-0600	68-0602	68-0603	68-0605	68-0606
68-0607	68-0608	68-0609	69-0350		

* RF-4C Modified to add AN/ASQ-203(V)1 Navigation and Weapon Delivery System
# 65-0904 Destroyed in a mishap on May 17, 1990
# 65-0929 Destroyed in a mishap on November 18, 1986

F-4G (Received from the 35 FW, George AFB, and the 52 FW, Spangdahlem AB)

69-0238	69-0243	69-0249	69-0261	69-0264	69-0265
69-0272	69-0273	69-0275	69-0277	69-0298	69-0303
69-0305	69-0306	69-0307	69-7202	69-7204	69-7207
69-7208	69-7217	69-7220	69-7256	69-7263	69-7272
69-7290	69-7294	69-7297	69-7298	69-7546	69-7551
69-7557	69-7560	69-7561	69-7572	69-7574	69-7580

Twenty RF-4Cs are seen here on the ramp at Gowen Field, Idaho in June 1989. Visible is the blue and gold IDAHO tail stripe. (David F. Brown Collection)

RF-4C 65-0929 is seen here on the ramp at Hill AFB, Utah on March 30, 1978 in standard Southeast Asia Camouflage. (Brian C. Rogers)

RF-4C 65-0845 is seen here on the ramp at Offutt AFB, Nebraska in July 1982 in Southeast Asia Wraparound Camouflage. (George Cockle)

RF-4C 68-0595 is seen here on the ramp at Davis-Monathan AFB, Arizona on March 31, 1984 in standard Southeast Asia Camouflage. As indicated by the towel rack antenna, 0595 is equipped with the AN/ARN-101 LORAN navigation system. (Brian C. Rogers)

RF-4C 65-0902, a NWDS equipped jet, is seen here on the ANG ramp at Phoenix Sky-Harbor on November 20, 1983 in Southeast Asia Wraparound Camouflage. The white 1 on the tail is from the PHOTO FINISH 83 meet. (Brian C. Rogers)

RF-4C 65-0904 is seen here in front of its hangar at Gowen Field, Idaho in October 1980 in Southeast Asia Wraparound Camouflage. The aircraft is fitted with the yellow vinyl covered canvas intake covers. The aircraft crashed on May 17, 1990, while assigned to the 120th TRG. (Don Logan Collection)

RF-4C 65-0874 is seen here in Europe I Camouflage taxiing for takeoff in October 1984. (David F. Brown)

RF-4C 65-0918 is seen here in Europe I Camouflage on April 4, 1986. (Kevin Patrick)

The crew of RF-4C 65-0606 is seen preparing for departure on September 3, 1987. As indicated by the towel rack antenna, this aircraft is equipped with the LORAN navigation system. (Brian C. Rogers)

RF-4C 65-0823 is seen here on the ramp in Hill Grays in September 1990. (Ben Knowles)

RF-4C 64-1063 is seen here on the ramp at Nellis AFB on April 19, 1990 in Hill Grays paint.Visible on the right underwing pylon is an Aircraft Instrumentation System (AIS) pod, part of the Air Combat Maneuvering Instrumentation (ACMI) system . (Brian C. Rogers)

RF-4C 65-0910 equipped with AN/ASQ-203 NWDS is seen here on the ramp at Nellis AFB in November 1990. (Ben Knowles)

RF-4C 68-0600 equipped with AN/ARN-101 LORAN navigation is seen here on August 7,1987. (Doug Remington)

RF-4C 65-0823 is seen here on the ramp at Gowen Field, Idaho on April 1992 with the new blue and white IDAHO tail stripe. The NWDS equipped jet was transferred to Spain as CR12-55. (Don Logan)

RF-4C 65-0910 is seen here on March 1992 with the blue and white IDAHO tail stripe. (Don Logan Collection)

F-4G 69-0298 with the distinctive Wild Weasel mouth and 69-0202 are seen here from the Copilots window of a B-52 over the Nellis Range Complex during a Red Flag Mission on February 29, 1996. (Brian C. Rogers)

F-4G 69-7551, the aircraft assigned to the author in 1972 as an F-4E of the 469th TFS, 388th TFW at Korat Royal Thai Air Force Base, is seen here taxiing out for a practice mission. Visible on the Triple Ejector Racks (TERs) under the wings are BDU- BDU-33 25-pound practice bombs. (Jerry Geer)

F-4G 69-0249, armed with an ATM-88B training missile, taxis to parking on the 184th TFW ramp at McConnell AFB in July 1992. The blue stripe on the missile between the guidance section and warhead section indicates the missile is not a "live" AGM-88 HARM, but rather a training missile with no warhead. 0249 was converted to QF-4G AF-163. (Don Logan)

F-4G 69-0303, carrying an ATM-88B, is seen here in July 1995 wearing the gray IDAHO tail stripe. 0303 was converted to QF-4G AF-154 and destroyed on October 21, 1998. (Don Logan Collection)

F-4G 69-0306, carrying an ATM-88B, is seen here in July 1994. 0306 was converted to QF-4G AF-131. (Jerry Geer)

F-4G 69-7220, carrying a training AIM-9 Sidewinder Air-to-Air missile, is seen here in April 1992 wearing the blue and white IDAHO tail stripe. 7220 was converted to QF-4G AF-150. (Don Logan)

F-4G 69-0298, wearing a standard F-4E type shark mouth, is seen here on April 18, 1996. It carries an IDAHO tail stripe applied in two shades of gray. (Bill Spidle)

To differentiate F-4G Wild Weasels from their F-4E Fighter-Bomber brothers, a Weasel mouth, like the one shown here on 69-0272 was developed by the 190th FS at Gowen Field. (Bob Greby)

F-4G 69-0272 seen here on March 2, 1996 wears the Weasel mouth applied by the FS. The 561st Fighter Squadron at Nellis AFB was the only other unit to carry the Weasel mouth. 0272 was converted to QF-4G AF-161. (Bob Greby)

F-4G 69-0305 seen here on February 8, 1996 carries WW (Wild Weasel) tail codes applied by the 190th FS Idaho ANG. The 35th Fighter wing at George AFB also wore WW tail codes on their F-4Gs. 0305 was converted to QF-4G AF-170 and destroyed on August 3, 2000. (Jack Callaway)

F-4G 69-0298 is seen here on February 8, 1996 on the Ramp at Gowen Field. 0298 was converted to QF-4G AF-149 and destroyed on May 13, 1998. (Jack Callaway)

F-4G 69-7272 seen here in September 1995 carries an unusual serial number presentation. The 7 was added to the 272 is because the unit also had 69-0272. 7272 was converted to QF-4G AF-158 and destroyed on October 29, 1998. (Ben Knowles)

F-4G 69-7202, carrying a training AIM-9 Sidewinder Air-to-Air missile, is seen here in September 1995 wearing WW tail codes and the gray IDAHO tail stripe. (Ben Knowles)

F-4G 69-7557, wearing a standard F-4E type shark mouth, is seen here on April 18, 1996. It carries an IDAHO tail stripe applied in a single shade of gray. 7557 was converted to QF-4G AF-159. (Jack Callaway)

F-4G 69-7551, a Combat veteran of both Southeast Asia and Desert Storm seen here in November 1998, was saved from the Aerial Target program and has been put on display at Gowen Field, Idaho. (Kevin Jackson)

170th TFS, 183rd TFG, Capital Airport, Springfield, Illinois

As part of the Illinois Air National Guard, the 170th TFS flew F-4C/Ds from Capital Airport in Springfield, Illinois and was organized in Springfield in September 1948. The 170th TFS 'Fly 'N Illini' has flown F-51Ds both from Springfield and while on active duty in Texas and California during the Korean War, then beginning in 1953 F-86Es and beginning in 1955 F-84Fs. During 1961-1962 the 170th was placed on active duty at their home base during the Berlin Crisis in 1961-62. In November 1971, the 170th TFS F-84Fs were grounded due to structural corrosion.

Two months later the 170th received F-4Cs and become the first Guard unit to fly Phantoms in the tactical fighter role. Between October 1973 and July 1975, the 170th TFS was also assigned a small group of RF-4Cs. F-4Ds from the 474th TFW at Nellis AFB replaced F-4Cs in early 1981, and in June 1982 the unit first deployed overseas with F-4Ds when it went to the United Kingdom for Coronet Bravo. F-4Ds of the 170th TFS wore **SI** (**S**pringfield **I**llinois) tail codes. Conversion from Phantoms to F-16A/Bs was officially completed on October 1, 1989.

Below: MiG killer F-4C 63-7699 seen here on April 12, 1975, was responsible for downing a North Vietnamese MiG-17 on May 14, 1967. (Bob Burgess)

183 TFG (IL ANG)

F-4C

63-7453	63-7490	63-7495	63-7505	63-7520	63-7523
63-7570	63-7602	63-7650*	63-7672	63-7683*	63-7699*
63-7707	64-0658	64-0679*	64-0691	64-0713	64-0757
64-0758	64-0766	64-0770	64-0776***	64-0811	64-0822
64-0825	64-0893	64-0899			

* 63-7650 - MiG-17 - September 16, 1966
* 63-7683 - MiG-21 - January 2, 1967
* 63-7699 - MiG-17 - May 14, 1967
* 64-0679 - MiG-17 - July 10, 1965
*** 64-0776 - MiG-21 - April 23, 1967 - two - MiG-21 - May 22, 1967

RF-4C

Six aircraft were assigned during the mid-1970s as proficiency trainers due to a shortage of available F-4C airplanes. They were transferred to 152nd TRW, Reno, Nevada as part of Nevada ANG's conversion to RF-4Cs.

65-0870	65-0875	65-0881	65-0886	65-0890	65-0897

F-4D (Received from the 474 TFW, Nellis AFB)

66-7505	66-7519	66-7536	66-7539	66-7545	66-7575
66-7579	66-7583	66-7587	66-7589	66-7593	66-7596
66-7604	66-7642	66-7648	66-7649	66-7652	66-7659
66-7663	66-7674	66-7675#	66-7696	66-7701	66-7705
66-7706	66-7725	66-7726	66-7730	66-7733	66-7742
66-7766	66-8705	66-8723	66-8808		

66-7675 Destroyed in a mishap on March 5, 1986

F-4C 64-0811 is seen here in May 1979 in the standard Southeast Asia Camouflage pattern and a red ILLINOIS tail cap. (Tom Brewer)

A flight of 170th TFS F-4Cs of the Illinois ANG are seen here in formation with 58-0068, a KC-135E of the 108th Air Refueling Squadron, Illinois ANG. (Norris Graser Collection)

The Three Illinois Air National Guard units, the 170th TFS flying F-4Cs from Springfield, the 108th Air Refueling Squadron flying KC-135Es from Chicago O'Hare International Airport, and the TFS flying OA-37 from Peoria fly in formation over Chicago. (Norris Graser Collection)

The same three Illinois Air National Guard aircraft fly formation over Lake Michigan. (Norris Graser Collection)

Triple MiG killer F-4C 64-0776 seen here on June 23, 1980, was responsible for downing three North Vietnamese MiG-21s; one on April 23, 1967, and two on May 22, 1967. (William R. Peake)

RF-4C 65-0870, seen here in front of Base Operations at Buckley ANGB, Colorado on August 1975, was one of the RF-4s used as pilot training aircraft before the 170 TFS received its F-4Cs in 1999. (Brian C. Rogers)

RF-4C's 65-0881 and 65-0897(whose tail is in the background) seen here were also used to train the 170th TFS pilots when the unit converted from F-100s to F-4s. (Brian C. Rogers)

RF-4C 65-0881 and 65-0897 are seen here at Springfield, Illinois on June 8, 1975. (Tom Brewer)

F-4C 63-7520 is seen here on October 7, 1980 in the standard Southeast Asia Camouflage pattern with a white ILLINOIS tail cap. (Jim Tunney)

Another view of F-4C 63-7520 seen here on June 1980. 7520 was written off on July 26, 1983 while assigned to the 142nd FIG, Oregon ANG. (Jerry Geer)

F-4C 63-7570 is seen here on May 7, 1980 in the standard Southeast Asia Camouflage pattern with a yellow ILLINOIS tail cap. (Robert J. Pickett)

Another view of F-4C 63-7570 seen here on October 7, 1980. 7570 was used as a battle damage repair aircraft at Robins AFB, Georgia and then scrapped. (Jim Tunney)

F-4D 66-7733 is seen here in August 1984 at Volk Field, Wisconsin in the Southeast Asia Wraparound Camouflage pattern and a yellow ILLINOIS tail cap. (James Goodall)

F-4D 66-7725 is seen here on August 8, 1984 in Europe 1 Camouflage pattern with a yellow ILLINOIS tail cap. An SP (SPringfield) tail code is almost invisible against the charcoal gray of the Europe 1 paint. (Charles B. Mayer)

F 4D 66 7723 is seen here on July 1987 in Europe I Camouflage pattern. The SI (Springfield, Illinois) tail code has been applied to the aircraft. (Bob Shane)

F-4D 66-7663 is seen here on July 1987 in Hill Grays. The tail markings are made up of a small SI tail code and non-standard five-digit serial number.(Bob Shane)

F-4D 66-7730 seen here on October 13, 1989 in Hill Grays has the standard TAC style SI tail code and serial number presentation. (Scott Van Aken)

163rd TFS, 122nd TFW, Baer Field, Fort Wayne, Indiana

The Indiana Air National Guard had two F-4E units, the 122nd TFW at Fort Wayne and the 181st TFG at Terre Haute. As the flying squadron of the 122nd TFW, the 163rd TFS was based at Fort Wayne Municipal Airport in Fort Wayne, Indiana. The two units after 1962 have had a parallel history. Between 1962 and 1964 the two units, although designated tactical fighter squadrons, were equipped with RF-84F Thunderflash reconnaissance aircraft. In 1979, they became the last ANG units to relinquish F-100Ds for Phantoms. The first Phantoms were F-4Cs including many modified for the Wild Weasel mission. However, neither the 163rd nor it sister squadron, the 113th at Terre Haute, ever flew their F-4Cs in the Weasel role, and in fact much of the Weasel equipment was removed from the airplanes upon transfer to the Guard. Both squadrons exchanged their F-4Cs for F-4Es in 1988, and were the only U.S. units to fly F-4Es equipped with the Boeing AN/ASQ-203(V)2 NWDS. The two Indiana squadrons were also the last ANG units to give up their F-4 fighters, leaving only RF-4Cs and F-4Gs in ANG service. Both squadrons began conversion to F-16C/Ds in the fall of 1991 and dropped 'Tactical' from their squadron designation in March 1992.

Organized in October 1947, the 163rd Fighter Squadron was initially equipped with F-51Ds, remaining at its home base while serving on active duty during the Korean War. The 163rd converted to F-80Cs in 1954, and beginning in March 1956 went on to fly F-86As, In January 1985 they converted to F-84Fs, being called up on active duty at Chambley AB, France, in 1961-1962. Beginning in May 1964, the 163rd was equipped with F-84Fs, replaced in June 1971 with F-100D/Fs. The first F-4C Phantoms arrived in 1979, coming from the 35th TFW at George AFB and the 52nd TFW at Spangdahlem AB, Germany. The F-4Cs were replaced with F-4Es from the 86thTFW, Ramstein AB, Germany in 1986. The 163rd began its conversion from F-4Es to F-16C/Ds in the fall of 1991 and was redesignated 163rd FS, 122nd FW, on March 16, 1992.

122 TFW (IN ANG)

F-4C (Received from the 39 TFTS, George AFB and the 81 TFS, Spangdahlem AB)

63-7428	63-7440 WW	63-7443	63-7452 WW	63-7467 WW	63-7474 WW
63-7481 WW#	63-7512	63-7513 WW	63-7545	63-7564 WW	63-7565
63-7567 WW	63-7594 WW	63-7596 WW	63-7615 WW	63-7644	64-0741 WW
64-0781 WW	64-0790 WW	64-0815 WW	64-0844	64-0847 WW	64-0922

WW F-4C Wild Weasel Modification
63-7481 Destroyed in a mishap on March 3, 1982

F-4E (Received from the 86 TFW, Ramstein AB)

67-0377*	67-0389*	68-0305*	68-0321*	68-0361*	68-0363*
68-0372*	68-0381*	68-0384	68-0393*	68-0394*	68-0402*
68-0404	68-0408*	68-0411*	68-0438*	68-0440*	68-0442*
68-0444*	68-0445*	68-0465#	68-0467*	68-0480*	68-0481*
68-0496*	68-0506*	68-0507	68-0509	68-0512	68-0515*
68-0516*	68-0517*	68-0528	68-0529*	68-0531*	68-0538*
69-7267					

* F-4E Modified to add AN/ASQ-203(V)2 Navigation and Weapon Delivery System
# 68-0465 Destroyed in a mishap on February 10, 1989

F-4C 64-0741, an ex-Wild weasel F-4C, is seen here at Nellis AFB on September 14, 1980 in the standard Southeast Asia Camouflage. This F-4C has a sharkmouth, highly unusual for a short nosed F-4 C/D Phantom II. The aircraft was sold to Tracor Flight Systems Inc. at Mojave, California and flew as N403FS. (Kirk Minert)

F-4E 68-0442 is seen here maneuvering in inverted flight. (Don Spering/AIR)

F-4E 68-0529 is seen in flight with F-16C 84-1377, both in 163rd FS markings. 0529 wears a custom paint job applied by Don Spering/AIR as part of the Fort Wayne Phabulous Phantom Pharewell. (Don Spering/AIR)

F-4E 68-0444 is seen here at McConnell AFB in April 1990, marked as the 122 TFW Flagship. The FW tail code comes from Fort Wayne, Indiana, the location of the 122 TFW. (Don Logan)

F-4E 68-0529 wears a custom paint job applied by Don Spering/AIR as part of the Fort Wayne Phabulous Phantom Pharewell on January 11, 1992. (Nate Leong)

This view of F-4E 68-0529 shows the special markings applied to the right side of the aircraft by Don Spering/AIR. (Nate Leong)

F-4C 63-7565 is seen here on August 25, 1984 in the Southeast Asia Wraparound Camouflage pattern. The rudder is marked indicating the aircraft has taken part in Operation CORONET CROWN. 7565 was later put on display at Hulman Field, Terre Haute, Indiana. (David F. Brown Collection)

F-4C 63-7443 is seen here on June 1980 in the standard Southeast Asia Camouflage pattern. As indicated by the small antenna on the fuselage just aft of the radome, 7443 had been modified as a F-4C Wild Weasel. (Jerry Geer)

F-4C 63-7452 is seen here in March 1980 in the standard Southeast Asia Camouflage pattern. 7443 was used as a battle damage repair aircraft at England AFB, Louisiana and then scrapped. (Douglas E. Slowiak/Vortex Photo Graphics)

F-4C 63-7596, an ex-Wild Weasel F-4C, is seen here on November 9, 1984 at Langley AFB in the Southeast Asia Wraparound Camouflage pattern. The intake splitter is marked with an F-16 silhouette, indicating an air-to-air victory over a Fighting Falcon. (David F. Brown Collection)

Above: Ex-Wild Weasel F-4C 63-7594 is seen here in May 1986 at Andrews AFB in the Europe I Camouflage. 7594 was used as a ground target at Harwood Air-to-Ground range, Wisconsin. (David F. Brown)

Right: The rudder on F-4E 68-0444, seen here in July 1987, indicates the aircraft has taken part in Operation CORONET CHEROKEE. 444, an NWDS jet, was transferred to the Greek Air Force under the Foreign Military Sales (FMS) program. (James Goodall)

Below: F-4E 68-0444 is seen here on taxiing, painted in the Southeast Asia Wrap-around Camouflage pattern. The rudder is marked indicating the aircraft has taken part in Operation CORONET CHEROKEE. (James Goodall)

F-4E 68-0531 seen here on September 26, 1989 in the Southeast Asia Wraparound Camouflage pattern was modified with the Boeing AN/ASQ-203 NWDS. (Don Logan Collection)

F-4E 68-0517 seen here on September 8, 1986 was also a NWDS jet and transferred to GREECE under the FMS program. (Douglas E. Slowiak/Vortex Photo Graphics)

F-4E 68-0372 seen here on May 10, 1988 in the Europe I Camouflage was also modified with the Boeing AN/ASQ-203 NWDS. (Phillip Huston)

NWDS equipped F-4E 68-0381 is seen here on August 18, 1981. Visible under the right wing above the empty TER rack is an AIM-9 Sidewinder captive training missile. 0381 was also transferred to GREECE under the FMS program. (David F. Brown)

After leaving the Indiana ANG, NWDS F-4E 68-0506, seen here in Hill Grays paint, was sold to GREECE under the Foreign Military Sales (FMS) program. (David F. Brown Collection)

113th TFS, 181st TFG, Hulman Field, Terre Haute, Indiana

As the flying squadron of the second of two Indiana ANG F-4 fighter units, the 113th TFS was based at Hulman Field Regional Airport in Terre Haute, Indiana. The first aviation unit in the Indiana Guard was organized as the 137th Squadron in 1921 and redesignated 113th Observation Squadron in 1923. During the World War II, it served on active duty in the U.S. until it disbanded in November 1943. Postwar, it flew F-51Ds, F-51Hs, F-80Cs, F-86As, F-84Fs, RF-84Fs and F-100D/Fs prior to converting to Phantoms.

The 113th received its F-4Cs in 1979, most coming from the 18th TFW at Kadena AB, Japan. The 113th TFS first deployed overseas when it went to Italy in May/June 1978 for Coronet Quail. In 1987 the 113th exchanged their F-4Cs for F-4Es from the 347th TFW, Moody AFB, Georgia. They were later modified with the Boeing NWDS. The 113th began conversion to F-16C/Ds in the summer of 1991 and dropped 'Tactical' from their squadron designation in March 1992.

181 TFG (IN ANG)

F-4C (Received from the 67 TFS, Kadena AB)

63-7418	63-7422	63-7423 WW	63-7433 WW	63-7437 WW	63-7440 WW
63-7443 WW	63-7447 WW	63-7452 WW	63-7459 WW	63-7470 WW#	63-7471
63-7473	63-7478 WW	63-7508 WW	63-7513 WW	63-7564 WW	63-7565 WW
63-7568	63-7574 WW	63-7581	63-7594 WW	63-7596 WW	63-7607 WW
63-7615 WW	63-7623★	63-7657	63-7665	64-0672	64-0675 WW
64-0724	64-0757 WW	64-0780	64-0781 WW	64-0783	64-0790 WW
64-0791 WW	64-0815 WW	64-0840 WW	64-0844 WW	64-0847 WW	

WW F-4C Wild Weasel Modification
★ 63-7623 - MiG-17 - May 20, 1967
63-7470 Destroyed in a mishap on November 19, 1986

F-4E (Received from the 347 TFW, Moody AFB & other bases)

67-0345*	67-0348*	67-0350*	67-0377*	68-0381*	67-0384*
67-0395	68-0318*	68-0326	68-0337*	68-0343	68-0346
68-0350	68-0361*	68-0405*	68-0412*	68-0418*	68-0420
68-0424*	68-0426*	68-0429#	68-0432*	68-0443*	68-0463
68-0478*	68-0490*	68-0495*	68-0503*		

* F-4E Modified to add AN/ASQ-203(V)2 Navigation and Weapon Delivery System
# 68-0429 Destroyed in a mishap on September 19, 1990

F-4Es of the 181st TFG are photographed on the Nellis AFB ramp on October 15, 1989 during Gunsmoke 89. The HF tail code stands for Hulman Field, Terre Haute, Indiana the home base of the 181st TFG (Marty J. Isham)

F-4E 68-0424 is seen here on April 6, 1991, marked as the 181st TFG Flagship. 0424, an AN/ASQ-203 NWDS jet, was transferred to the Greek Air Force under the Foreign Military Sales (FMS) program. (David F. Brown- top and Nate Leong bottom)

F-4E 67-0384 wears a custom paint job applied by Don Spering/AIR as part of the Terre Haute Phabulous Phantom close out on September 27, 1991 (Don Spering/AIR)

Above and right: F-4E 68-0384 wears a custom paint job applied by Don Spering/AIR as part of the Terre Haute Phabulous Phantom Pharwell on September 27, 1991. (Nate Leong)

F-4E 68-0384 is seen here on the ramp at Terre Haute with the Don Spering/AIR markings added for the Terre Haute Phabulous Phantom Pharwell on September 27, 1991. (Nate Leong)

F-4E 67-0384 in the Pharwell markings prepares to connect to the boom of a KC-135 during air refueling. (Don Spering)

F-4C 63-7623 is seen here on the ramp at Terre Haute on August 23, 1983. While assigned to the 8th TFW at Ubon RTAFB in Thailand, 7623 was responsible for the downing of a North Vietnamese MiG-17 on May 20, 1967. (William R. Peake)

Ex-Wild Weasel F-4C 63-7478 is seen here at Hulman Field on October 7, 1980 in the standard Southeast Asia Camouflage pattern. It was later used as a battle damage repair aircraft at Wright Patterson AFB, Ohio. (Jim Tunney)

F-4C 64-0675, an ex-Wild Weasel, is seen here in February 1987 in the Southeast Asia Wraparound Camouflage pattern. The rudder is marked indicating the aircraft has taken part in Operation CORONET PAWNEE NORWAY 86. 0675 was later used as a battle damage repair aircraft at Moody AFB, Georgia. (James Goodall)

Ex-Wild Weasel F-4C 63-7574 is seen here on February 28, 1983 at Luke AFB in the Southeast Asia Wraparound Camouflage pattern. (Douglas E. Slowiak/Vortex Photo Graphics)

F-4C 63-7437, an ex-Wild Weasel aircraft, is seen here on September 21, 1985 in the Europe I Camouflage. The inscription on the nose commemorates 5000 flight hours for Lt. Col Wayne Yarolem. Lt. Col Yarolem was the high time USAF F-4 pilot. During his 24 years in the Air Force from July 7, 1967 until June 8, 1991 Lt. Col. logged a total of 6514.6 hours in the F-4. This included 392 combat sorties and 776 combat hours. 7437 was later used as a battle damage repair aircraft at Eglin AFB, Florida. (Douglas E. Slowiak/Vortex Photo Graphics)

F-4C 63-7418 is seen here on May 17, 1985 in the Europe I Camouflage. A "ZAP" from the 140th TFG Cougars of the Colorado ANG is visible on the intake. (William R. Peake)

F-4C 63-7423 is seen here on June 29, 1987 in the Europe I Camouflage, also marked for Operation CORONET PAWNEE NORWAY 86. 7423, an ex-Wild Weasel, was put on display at Koblenz-Lutzel, Germany on May 30, 1991. (Douglas E. Slowiak/Vortex Photo Graphics)

F-4C 63-7447 is seen here on July 1987 in the Europe I Camouflage also marked for Operation CORONET PAWNEE NORWAY 86. 7447, an ex-Wild Weasel, was used as a target on the Nellis AFB ranges. (Bob Shane)

Ex-Wild Weasel F-4C 63-7607 is seen here in July 1987 in the Hill Grays Camouflage also marked for operation CORONET PAWNEE NORWAY 86. Aircraft in three different camouflage schemes (Southeast Asia Wraparound, Europe I, and Hill Grays) all from the 181st TFG took part in Operation CORONET PAWNEE NORWAY 86. 7607 flew for Tracor Flight Systems as N423FS before being scrapped. (Bob Shane)

F-4E 68-0412 is seen on takeoff at Nellis AFB in October 1989 during Gunsmoke 89. The 181st Gunsmoke 85 emblem is visible on the intake. 0412, an AN/ASQ-203 NWDS jet, was transferred to the Greek Air Force under the Foreign Military Sales (FMS) program. (David F. Brown)

F-4E 68-0348 is seen on taxiing at Lambert Field, St Louis, Missouri on September 13, 1991. 0348, an AN/ASQ-203 NWDS jet, was transferred to the Turkish Air Force under the Foreign Military Sales (FMS) program. (Keith Snyder)

F-4E 68-0478 is seen on July 26, 1991. (Norris Graser)

127th TFS and 177th TFTS, 184th TFG, McConnell AFB, Wichita, Kansas

The 184th Tactical Fighter Group, Kansas ANG was based at McConnell AFB, Kansas, with both the 127th TFS and 177th TFTS flying F-4Ds.

The 127th TFS was activated as part of the Kansas National Guard in August 1941, the 127th Observation Squadron was introduced into federal service two months later. During World War II the unit was ordered to extended active duty in the U.S. Army Air Corps until October 6, 1945. On September 7, 1946 the unit was redesignated the 127th FS and began flying F-51D Mustangs. It converted to F-84Cs in 1950. May 4, 1952, after joining the 137th Fighter Bomber Wing at Alexandria, Louisiana, it moved to Chaumont AB, France. The 127th Fighter-Bomber Squadron returned to Wichita, Kansas 21 months later. Due to a shortage of jets brought about by the conflict in Korea, the 127th re-equipped with F-51Ds. On June 10, 1954 the 127th returned to jets when it began flying F-80Cs. On January 10, 1958 the unit was redesignated as the 127th Fighter Interceptor Squadron and converted to F-86Ls flying in an Air Defense role.

Converting to F-100Cs on March 25, 1961, the 127th was again redesignated, changing to a TFS. In January 1968, following the North Korean seizure of the USS Pueblo, the 127th TFS was called to active duty and deployed to Kunsan AB, Korea. It returned to state control in June 1969. On March 25, 1971, the squadron converted to F-105D/Fs and, as the 127th TFTS, became the Combat Crew Training School conducting F-105 pilot training.

On August 7, 1979, the 184th received its first F-4D Phantom II. Many of its initial F-4Ds came from RAF Bentwaters in the Untied Kingdom. On October 8, 1979, the unit was redesignated as the 184th Tactical Fighter Group and was tasked to provide F-4 crew training for the ANG. In April 1982, the 184th was tasked with developing an F-4 Fighter Weapons Instructor Course to meet the needs of the Air Force, Air Force Reserve, and Air National Guard. The 184th received additional jets from the 18th TFW at Kadena AB, Japan, the 56th TFW at MacDill AFB and 474th TFW at Nellis AFB in 1980 through 1982. In addition, the 134th Tactical Control Flight was established at McConnell AFB on September 15, 1982, to provide tactical radar control for the increased training.

Additional F-4Ds were added from the 116th TFW, Georgia ANG in 1983 and 1984. On February 8, 1984, a second flying squadron, the 177th Tactical Fighter Training Squadron was established joining the 127th TFS.

The responsibility of the 177th was for the conversion and upgrade training of all aircrew in the F-4. In August 1985, the 127 TFS took over the F-4 FWIC mission from the regular Air Force, concurrent with inactivation of the F-4 Division, USAF Fighter Weapons School at Nellis AFB.

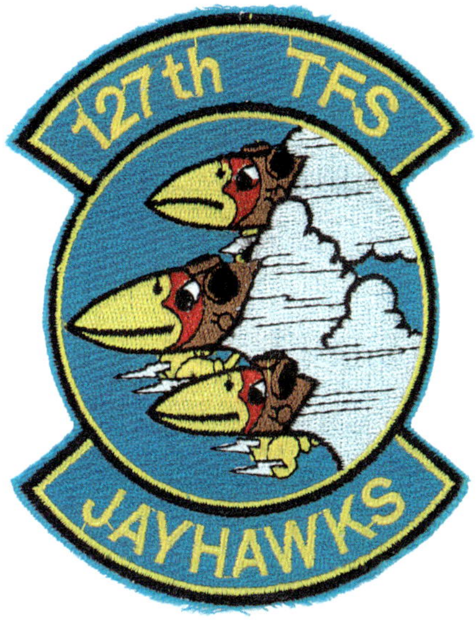

To meet the F-4 training requirements, the 184th Tactical Fighter Group with 70 F-4Ds assigned achieved an annual flying program of 9600 sorties. In August 1985, the unit reached its first 1000-sortie month. In January 1987, the 184th began to convert to the F-16A/B Fighting Falcon. With this new aircraft, the unit was tasked with conducting conversion and upgrade training for the ANG. On July 8, 1987, the 161st Tactical Fighter Training Squadron was established as the third flying squadron of the 184th Tactical Fighter Group. Formal activation ceremonies for the 161st occurred on September 12, 1987, with the unit operating ten F-16s and conducting its first student training class. In August 1988, the 127th Tactical Fighter Squadron graduated its final class from the F-4 Fighter Weapons Instructor Course. Following the closing of this course, the 127th became the second squadron to convert to the F-16. With the departure of the last F-4D on March 31, 1990, the 177th became the last squadron to convert to the F-16.

In July 1994, the 184th Fighter Group was designated at the 184th Bomb Group flying the B-1B Lancer. The 184th Bomb Group was redesignated the 184th Bomb Wing on October 1, 1996. The 184th was the first Air National Guard unit to fly bombers.

184 TFG (KS ANG)

F-4D

65-0590	65-0595	65-0597	65-0609	65-0614	65-0615
65-0638	65-0641	65-0643	65-0644	65-0660	65-0666
65-0680	65-0681	65-0687	65-0692	65-0701	65-0702
65-0703	65-0705	65-0710	65-0711	65-0714	65-0736
65-0741	65-0749	65-0754	65-0756	65-0759	65-0769
67-0772	65-0778	65-0782	65-0783	65-0793	65-0794
65-0798	66-0229	66-0234	66-0239★	66-0242	66-0259
66-0261	66-0268★	66-0271★★	66-0274★	66-7459	65-7461
66-7475	66-7479	66-7484	66-7486	66-7502	66-7520
66-7539	66-7553	66-7566	66-7579	66-7588	66-7595
66-7604	66-7605	66-7614	66-7633	66-7662	66-7663
66-7666#	66-7668	66-7676	66-7681	66-7689	66-7694
66-7714	66-7720	66-7721	66-7735	66-7746	66-7759
66-7768	66-8683	66-8693	66-8783	66-8815	

★ 66-0239 - MiG-21 - September 9, 1972
★ 66-0268 - MiG-21 - October 12, 1972
★★ 66-0271 - MiG-21 - July 18, 1972 - MiG-21 - July 29, 1972
★ 66-0274 - MiG-17 - October 26, 1967
66-7666 Destroyed in a mishap on July 1, 1985

F-4D 66-0271, a double MiG killer is seen here in flight. 0271, while assigned to the 432nd TFR was credited with downing two North Vietnamese MiG-21s, one on July 18, 1972 and the second on July 29, 1972. (KS ANG)

F-4D 65-0666 is seen here in flight over the South West U.S. with the Northrop F-20 demonstrator with an ANG emblem on the tail and an Arizona A-7D in a special test camouflage. (KS ANG)

F-4D 66-7461 is seen here in front of the 184th TFTG Hangar at McConnell AFB. It is carrying a special (not certified) load of five SUU-21 20mm gun pods and four Aim-9 Sidewinder Missiles. The SUU-21 on the centerline is not visible in the photo. (KS ANG)

F-4D 66-7484 in Hill Grays and 66-0229 in Europe 1 are seen here in front of the 184th TFTG Hangar at McConnell AFB. (KS ANG)

F-4D 66-0229 is seen here in flight in March 1981 painted in the standard Southeast Asia Camouflage scheme. (Jerry Geer)

This 184th TFTG F-4D pulls up on the wing during low level flight over Kansas on AUGUST 1988. (Don Spering/AIR.)

Covered with signatures of 184th TFTG Members 66-7553, the 184th's last F-4 prepares for departure to AMARC on April 1990 as part of the transition of the 184th from F-4s to F-16s. (Don Logan)

F-4D 66-0271, the 184th's double MiG killer taxis past the ramp at McConnell AFB on July 7, 1988. (Brian C. Rogers)

The double MiG killer stars are visible on the right intake ramp of F-4D 66-0271. 0271 was repainted into Southeast Asia Camouflage and remains on display at the 184th Facility at McConnell AFB, Wichita, Kansas. (Don Linn)

F-4D 66-0268, a MiG killer is seen here in Europe 1 Camouflage. (Marty Isham Collection)

F-4D 65-0595 is seen here on April 19, 1980 on the McConnell flight line painted in the standard Southeast Asia Camouflage scheme with full color 184th TFTG Markings. (Don Logan)

F-4D 65-0595 is seen here on July 26, 1980 on the flight line with Happy Hooligans F-4Ds of the North Dakota Air National Guard. (Douglas E. Slowiak/Vortex Photo Graphics)

F-4D 65-0666 is seen here on the flight line at McConnell AFB painted in the standard Southeast Asia Camouflage scheme. The 184th also had F-4D 66-0666 assigned (Don Logan)

F-4D 66-7681 is seen here at Luke AFB painted in the Southeast Asia Wraparound Camouflage scheme and low visibility unit emblems. The four digit tail number was applied to differentiate 66-7681 from 65-0681 also operated by the 184th TFG. (Douglas E. Slowiak/Vortex Photo Graphics)

F-4D 65-0736 is seen here on August 6, 1983, also at Luke AFB painted in the Southeast Asia Wraparound Camouflage scheme and low visibility unit emblems. (Douglas E. Slowiak/Vortex Photo Graphics)

F-4D 65-0759 is seen here on August 29, 1983, painted in the Southeast Asia Wraparound Camouflage scheme. (David F. Brown Collection)

F-4D 65-0579 is seen here painted in the Southeast Asia Wraparound Camouflage scheme. (Paul Osborne)

F-4D 65-0590 is seen here on the 184th ramp at McConnell AFB on March 23, 1986 painted in the Gloss Gray scheme with full color markings and a light gray radome. (Tom Brewer Collection)

F-4D 66-7668 is seen here on the 184th ramp at McConnell AFB painted in the Gloss Gray scheme with full color markings with a black radome. (Tom Brewer Collection)

F-4D 65-0714 is seen here on the 184th ramp at McConnell AFB painted in the Gloss Gray scheme with full color markings. (Tom Brewer Collection)

F-4D 65-0756 is seen here taxiing at McConnell on March 22, 1988, painted in the Europe I Camouflage scheme. (Brian C. Rogers)

F-4D 66-0229 is seen here in the Europe I Camouflage scheme. A camel kill marking is visible on the intake splitter. (William R. Peake)

F-4D 66-7694 is seen here at Nellis AFB on April 28, 1988, painted in the Hill Grays scheme with low visibility unit markings. The Air National Guard Bureau General McMerty, (a former Commander of the 184th TFTG) has his name painted on the nose. 7694 was later used as a target on the Avon Park Gunnery Range in Florida. (Marty Isham)

F-4D 65-0705 is seen here at Nellis AFB on July 18, 1987, painted in the Hill Grays scheme with low visibility unit markings. 0705 was moved to Holloman AFB and placed on the range as a target. (Don Logan Collection)

Nose gear door art on 65-0782 (Don Logan)

Nose gear door art on 66-7484 (Don Logan)

Nose gear door art on 66-7553 (Don Logan)

Nose gear door art on 66-7694 (Don Logan)

Nose gear door art on 66-7702 (Don Logan)

Nose gear door art on 66-7710 (Don Logan)

Nose gear door art on 66-7756 (Don Logan)

Nose gear door art on 66-7768 (Don Logan)

Nose gear door art on 66-8783 (Don Logan)

165th TRS, 123rd TRW, Standiford Field, Louisville, Kentucky

Based at Standiford Field, Louisville, the 165th TRS received its first RF-4C in February 1976. The 165th Fighter Squadron (SE) was organized in February 1947 and flew North American F-51Ds. It was called to active duty. In November 1951, and moved, while still on active duty, to England, where it flew F-84Es from RAF Manston. Returning to Louisville and state control in July 1952, the 165th Fighter-Bomber Squadron was equipped with F-51Hs and then returned to F-51Ds. The 165th became a fighter interceptor squadron in July 1955, flying the T-28A trainer for a few months until ready for F-86As.

The squadron became a Tactical Reconnaissance unit in January 1958, flying RB-57Bs, RF-84Fs, RF-101G/Hs (including active duty in 1968-1969 when it operated from Richards-Gebaur AFB, Missouri, and from Itazuke AB, Japan) and RF-101Cs before converting to RF-4Cs in 1976. A KY tail code was carried on RF-4Cs of 123rd TRW. The unit first deployed its RF-4Cs to Europe in February/March 1978 when it went to Norway as part of Coronet Snipe. On 8 January 1989, the unit was redesignated 165th TAS, 123rd TAW and replaced its RF-4Cs with Lockheed C-130Bs. Redesignated 165th AS, 123rd AW, in March 1992, the Kentucky ANG unit is now equipped with new C-130Hs.

123 TRW (KY ANG)

RF-4C

64-1031	64-1035	64-1060	64-1061	64-1063	64-1067
64-1069	64-1070	64-1071	64-1073[#]	64-1074	64-1075
64-1080	64-1081	64-1083	64-1084	65-0822	65-0823
65-0826	65-0835	65-0836	65-0837	65-0841	65-0851
65-0852	65-0873	65-0944			

[#] 64-1073 Destroyed in a mishap on May 16, 1992

RF-4C 64-1063 is seen in August 1982 here with the KY (KentuckY) tail code painted in Southeast Asia Wraparound Camouflage paint. (Don Spering/AIR)

RF-4C 64-1084 is seen here in December 1978 in standard Southeast Asia Camouflage paint. The Air National Guard Bi-Centennial Emblem is on the nose. The aircraft was lost on October 13, 1989 while assigned to the 67th TRW, Bergstrom AFB, Texas. (David F. Brown Collection)

RF-4C 64-1070 is seen here taxiing on June 21, 1980 painted in standard Southeast Asia Camouflage paint. 1070 was transferred to Spain under the FMS program as CR12-46 and modified with AN/ASQ-203 NWDS. (Ben Knowles)

RF-4C 64-1067 is seen here on the ramp with the serial number in the early white and light gray configuration painted on the tail (David F. Brown Collection)

RF-4C 64-1063 is seen here on the ramp in July 1982 painted in Southeast Asia Camouflage Wraparound paint. 1063 was later modified with AN/ASQ-203 NWDS. (Ben Knowles)

RF-4C 64-1081 is seen here on the ramp in May 1982 painted in Southeast Asia Camouflage Wraparound paint and black KY tail code. The rudder is painted for PHOTO DERBY 82 competition. 1081 remained at Louisville and has been put on display. (David F. Brown Collection)

RF-4C 64-1071 is seen here on the ramp in May 1983 painted in Southeast Asia Camouflage Wraparound paint and black KY tail code. The rudder is painted for CORNET BISHOP 83 exercise. (Tom Brewer Collection)

Right: The flying horse of the 123rd TRW is seen here on the tail of RF-4C 65-0837 in November 1986, specially marked for RAM 86. (Tom Brewer Collection)

Below: 65-0823 also represented the 123rd TRW at RAM 86. (Tom Brewer Collection)

RF-4C 64-1075 is seen here on the ramp in June 1982 painted in Southeast Asia Camouflage Wraparound paint and black KY tail code. Markings for NATO BEST FOCUS 82 competition are painted on the intake and rudder. (Ted Van Geffen)

RF-4C 64-1084 is seen here on the ramp in June 1986 painted in Europe 1 paint and black KY tail code. (David F. Brown Collection)

RF-4C 64-1075 is seen here on the ramp in April 1984 painted in Europe 1 paint. A "Zap" in the shape of the state of Idaho, applied by the 124th TRG at Boise is visible on the intake splitter. (David F. Brown)

RF-4C 65-0841 is seen here on the ramp in July 1987 painted in Hill Grays. The flying horse has been added to the white Kentucky tail cap. 0841 was transferred to Spain under the FMS program as CR12-50 and modified with AN/ASQ-203 NWDS. (David F. Brown)

RF-4C 64-1074 is seen here taxiing in July 1988 painted in Hill Grays. The tail cap with the flying horse and KENTUCKY script is now painted subdued gray. The words PHANTOMS FINEST have been added to the intake. (David F. Brown)

RF-4C 65-0873 is seen here on the ramp in July 1988 painted in Hill Grays with PHANTOMS FINEST applied to the intake. 0873 was transferred to Spain under the FMS program as CR12-52 and modified with AN/ASQ-203 NWDS. (David F. Brown)

122nd TFS, 159th TFG, NAS New Orleans, Louisiana

The 122nd TFS, Louisiana ANG was based at NAS New Orleans, flying F-4Cs between the spring of 1979 and the summer of 1985. It was organized in March 1941 as the 122nd Observation Squadron, and was called to active duty seven months later. It flew ASW missions in the United States and North Africa before becoming an RTU for American and French pilots in Algeria. Equipped with B-17s and B-24s, it was redesignated the 885th Bombardment Squadron in May 1994. As a special unit of the Fifteenth Air Force it flew partisan support missions in the Balkans. Following World War II, the Louisiana unit first flew A-26s as the 122nd Bombardment Squadron. As an air defense squadron, the 122nd FIS briefly flew F-80Cs in 1957, prior to converting to F-86Ls later in that year and to F-102A/TF-102As in 1960.

In July 1970 the 122 converted to a Tactical Fighter Squadron flying F-100D/Fs for nine years until converting to F-4Cs in 1979. The aircraft came from George AFB, Luke AFB, and Torrejon AB. The unit's F-4Cs did not carry tail codes, but rather had Coonass Militia applied to the top or leading edge of the vertical tail for a short period circa 1982-83. The 122nd TFS took its F-4Cs to Iceland for Coronet Rodeo in August/September 1982. It converted to McDonnell Douglas F-15A/Bs in 1985. The 122nd FS is now known as the Bayou Militia; the term Coonass Militia being deemed politically incorrect (a Coonass being a native of the southern part of Louisiana).

Some F-4Cs of the 122nd TFS carried one of two special two gray camouflage schemes. These schemes were not applied to any other unit's aircraft.

159 TFG (LA ANG)

F-4C (Received from George AFB, Luke AFB, and Torrejon AB)

63-7417	63-7448	63-7463	63-7471	63-7492#	63-7506
63-7522	63-7552	63-7555	63-7556	63-7631	63-7637
63-7679	63-7704*	64-0659	64-0673*	64-0686#	64-0783
64-0794	64-0802	64-0823	64-0841	64-0852	

* 63-7704 - MiG-17 - May 14, 1967
* 64-0673 - MiG-17 - May 20, 1967
\# 63-7492 Destroyed in a mishap on January 14, 1984
\# 64-0686 Destroyed in a mishap on February 17, 1984

F-4C 64-0686 is seen here on September 1988 painted in standard Southeast Asia camouflage scheme with a black serial number and white COONASS MILITIA on the tail. (Don Spering/AIR)

F-4C MiG killer 63-7704 seen here at NAS New Orleans on May 5, 1983 was credited with the destruction of a North Vietnamese MiG-17 shot down on May 14, 1967 while the aircraft was assigned to the 408th TFS, 366th TFW. The aircraft is painted in the first of the special camouflage schemes used by the COONASS Militia. COONASS is slang for a person from Southern Louisiana. 7704 was placed on display in Milwaukee, Wisconsin. (Scott Wilson)

F-4C 63-7552 is seen here on the ramp on August 17, 1984 painted in Southeast Asia Wraparound camouflage scheme. It has a large red crawfish painted on the intake splitter. 7552 was used as a battle damage repair aircraft at Bergstrom AFB, Texas and later scrapped. (David F. Brown Collection)

F-4C 64-0852 is seen here on the ramp in April 1980 painted in standard Southeast Asia camouflage scheme with white tail markings. (Frank MacSorley)

F-4C 63-7555 is seen here on the ramp in July 1980 painted in standard Southeast Asia camouflage scheme with black tail markings. (David F. Brown Collection)

F-4C 63-7555 is seen here on takeoff at NAS New Orleans on May 20, 1983. 7555 was placed on display at the Yankee Air Museum, Michigan. (Ray Leader)

F-4C 63-7556 is seen here on the ramp on March 12, 1982 painted in standard Southeast Asia camouflage scheme with black tail markings. Of note are the four aircraft carrier kill markings on the intake splitter. 7556 was placed on display at the Louisiana Military History Museum. (Douglas E. Slowiak/Vortex Photo Graphics)

F-4C 63-7506 is seen here on the ramp on May 16, 1985 painted in Europe I camouflage scheme with black tail markings. 7506 was placed on the Nellis AFB Range as a target. (William R. Peake)

F-4C 64-0852 seen here in September 1983 painted in a third special camouflage scheme with COONASS MILITIA in black script on the tail. 0852 was placed on the Holloman AFB Range as a target. (Bob Shane)

F-4C 63-7637 seen here on May 4, 1983 painted in the fourth special camouflage scheme with COONASS MILITIA is white script on the tail. (Charles B. Mayer)

F-4C 63-7463 is seen here on the ramp at Nellis AFB on April 2, 1984 painted in a third special camouflage scheme used by the Coonass Militia. 7463 was placed on display at the Camp Robinson Museum in Arkansas. (Keith Snyder)

F-4C MiG killer 63-7704 is seen in the special camouflage scheme in May 1983 with COONASS MILITIA in white script on the tail. (Scott Wilson)

171st FIS, 191st FIG, Selfridge Air National Guard Base, Michigan

The 171st FIS was located at Selfridge Air National Guard Base on the north side of Detroit, Michigan. The 171st FIS was the first ANG air defense unit to fly Phantoms, converting from F-106A/Bs to F-4Cs at Selfridge ANGB in the spring of 1978. The 107th Fighter Squadron (SE) was extended federal recognition in April 1948, and flew F-51Ds and F-84Bs in the tactical role before converting in October 1958 to F-51Hs and becoming the 171st FIS. As an FIS, the 171st flew F-86Es and F-89Cs. In February 1958 it was redesignated the 171st TRS flying RF-84Fs followed by RF-101Cs. In January 1971 it was once again designated 171st FIS fling F-106A/Bs.

In 1978 the squadron converted to F-4Cs from various sources including Keflavik, Iceland, Tyndall AFB, Florida, and a few from the Illinois ANG. Maintaining an alert detachment at Seymour Johnson AFB, North Carolina, the squadron traded its F-4Cs for F-4Ds during the summer of 1986. The F-4Ds were transferred from the 158th TFG, Burlington IAP, Vermont ANG. In 1988 the 171st FIS completed its conversion to Fighting Falcons, with its F-16A/Bs subsequently being upgraded to the ADF configuration, on July 1, 1990.

191 FIG (MI ANG)

F-4C (Received from Keflavik, Iceland and Tyndall AFB)

63-7412 – 01	63-7436 – 02	63-7442 – 03	63-7446 – 04	63-7453 – 05	63-7460 - 06
63-7475 – 07	63-7482 – 08	63-7514 – 08	63-7529-10/09	63-7534 – 10	63-7536
63-7576-13/12	63-7583 – 13	63-7589 – 14*	63-7595 – 15	63 7610 – 16	63-7618 - 15
63-7622-18/16	63-7626	63-7666 – 18	63-7679	63-7685 – 19	64-0706
64-0707 - 20					

The 2-digit numbers after the F-4C serial numbers are the NORAD tactical numbers as painted on the aft fuselage sides of the airplanes from 1978 until around 1984. Most aircraft were re-numbered once.

* 63-7589 - MiG-21 - January 2, 1967

F-4D (Received from Vermont ANG)

65-0590	65-0598	65-0611	65-0617	65-0646	65-0667
65-0688	65-0694	65 0701	65 0714	65-0730	65-0734
65-0737	65-0765	65-0774	65-0775	65-0779	65-0788
65-0790	65-0793	65-0794	66-0226	66-0243	66-0244
66-0254	66-0266	66-0279	66-0283	66-7457	66-7464
66-7470	66-7471	66-7486			

Michigan F-4C 63-7482 fires an AIM-7 Sparrow missile from the right forward Missile well. 7482 was put on display at the Minnesota ANG Historical Museum. (Boeing)

An AIM-7 Sparrow is visible in the right forward Missile well of F-4C 63-7482. (Boeing)

Three Michigan F-4s, One F-4C and two F-4Ds are seen here in flight in May 1986. (Doug Barbier)

Three Michigan F-4Ds, including the 191st FIG Commanders aircraft are seen here in flight in February 1989. (Doug Barbier)

F-4C 63-7685 is seen here in April 1982 escorting B-52G 58-0257 of the 319th Bomb Wing based at Grand Forks North Dakota. 7685 was later used as a target on the Nellis AFB Range Complex. (Doug Barbier)

F-4D 65-0737 is seen here in June 1988 escorting B-52H 60-0026 low level over the upper peninsula of Michigan. (Doug Barbier)

F-4D 66-0226, marked as the 171st FIS Commander's aircraft, is seen flying clean except for a single AIM-9 sidewinder on the right wing pylon. (Doug Barbier)

F-4C 63-7536, with CYRANO THE FEARLESS art on the intake, taxis out for a practice mission. After being used as a battle damage repair aircraft at Norvienich Air Base, Germany, 7536 was put on display at Seifertshofen, Germany. (James Goodall)

F-4C 63-7622 is seen here on the ramp in June 1988 painted in standard Southeast Asia camouflage scheme with Michigan tail markings. Of note is the numeral 16, the Air Defense ID Number, in white on the rear fuselage. 7622 was used as a battle damage repair aircraft at Shaw AFB, South Carolina. (Doug Barbier)

F-4C 63-7610, in Air Defense Gloss Gray is seen herein June 1983 with the 191st typical yellow and black checkerboard applied to the rudder and intake splitter. 7610 was used as a battle damage repair aircraft at RAF Lakenheath, England and then scrapped. (Doug Barbier)

F-4C 63-7666, in Air Defense Gloss Gray is seen here on the ramp at Offutt AFB, Nebraska in July 1984. 7666 was used as a battle damage repair aircraft at Kunsan Air Base, Republic of Korea. (Doug Barbier)

F-4C 63-7514, SHADOW DEMON, is seen here in May 1985. 7512 was used as a battle damage repair aircraft at Osan Air Base, Republic of Korea. (Doug Barbier)

Three 191st F-4s are completing their last checks prior to taking the runway for takeoff during June 1986. (Doug Barbier)

Two F-4Ds, one in full color markings and one in subdued grays taxi out for takeoff at McConnell AFB in April 1990. (Don Logan)

F-4D 65-0794, in Europe I is seen here in September 1986 with the 191st typical yellow and black checkerboard applied to the intake splitter. (Doug Barbier)

F-4D 66-7471, in Europe I is seen here on July 18, 1987 with a different style script MICHIGAN in yellow on the tail. (David F. Brown)

F-4D 65-0688 in Hill Grays seen here in July 1986 with only the script MICHIGAN in subdued gray on the tail. (David F. Brown Collection)

F-4D 65-0734 in Air Defense Gloss Grays seen here in September 1985 with only the script MICHIGAN in yellow gray on the tail. (Doug Barbier)

F-4D 65-0688 is seen here in full subdued gray markings on July 17, 1987. (David F. Brown Collection)

F-4D 66-0266 is seen here in modified subdued gray markings in September 1988. (David F. Brown)

The gloss gray paint of the Air Defense F-4s was perfect for the application of nose art. As a result, F-4s of the 191st FIG carried nose art starting with William Tell 82 in October 1982 and ending at the end of 1986. NEVER TRUST A SMILING CAT nose art on 63-7583 was carried as nose art on 63-7583 from October 1983 until the end of 1986. (Doug Barbier)

I DON'T TAKE DEFEAT LIGHTLY was nose art on 64-0707 from 1982 through 1984. (Doug Barbier)

In 1984, after the aircraft returned from Depot Maintenance, DON"T MESS WITH THE KID replaced I DON'T TAKE DEFEAT LIGHTLY as nose art on 64-0707. (Doug Barbier)

TRUSST ME was the nose art on 63-7529 for William Tell 82. (Doug Barbier)

DOUBLE TROUBLE was nose art carried on 663-7618 from 1983 through 1986. (Doug Barbier)

PUFF THE MAGIC DRAGON was the nose art on 63-7460 for William Tell 82 and remained on the aircraft through 1985. (Doug Barbier)

SHADOW DEMON nose art was carried on 63-7514 from December 1985 through 1986. SHADOW DEMON had been previously applied to 63-7442. (Doug Barbier)

BABY! replaced SHADOW DEMON as nose art on 63-7442 for William Tell 84 and remained on the aircraft through 1986. (Doug Barbier)

WE BAD was nose art on 63-7412 in October 15, 1984. The art was added for William Tell 84 staying until the aircraft went to Depot Maintenance in October 1985. (Don Logan Collection)

MAKE MY DAY was nose art on 63-7595 from William Tell 84 until it was repainted as part of programmed Depot Maintenance. (Doug Barbier)

THIS ONE'S FOR YOU BABY nose art was applied to 63-7666 for William Tell 84 and remained on the aircraft through 1986. (Don Logan Collection)

PATIENCE MY A** was nose art on 63-7482 from William Tell 82 through 1985. (Doug Barbier)

179th TRS/179th FIS, 148th TRG/148th FIG, Duluth International Airport, Minnesota

The 179th FIS operated from Duluth International Airport, located in Northern Minnesota at the western end of Lake Superior. The 179th Fighter Squadron (SE) was extended federal recognition in September 1948 with F-51Ds assigned. Flying in an Air Defense role, the 179th converted to F-94A/Bs in July 1954. Rocket-armed F-94Cs replaced gun-armed F-94A/Bs in Apri1 1957. These were replaced by F-89Js in July 1959. Convair F-102A/TF-102A Delta Daggers replaced them in November 1966. The 179th FIS converted to McDonnell F-101B/F Voodoos in April 1971.

The 179th became a tactical reconnaissance squadron during January 1976 when the 179th FIS finished conversion to RF-4Cs. In August 1979, in its first RF-4C deployment the 179th TRS took part in Coronet Bridle flying out of Germany.

In the fall of 1983, the unit converted to F-4Ds and on November 15, 1983 was once again designated the 179th FIS. Most of the F-4Ds were transferred from Homestead AFB. In 1986-87, it contributed F-4Ds, crews and maintenance personnel to Operation Creek Klaxon and assumed an air defense commitment at Ramstein AFB, Germany. During the winter of 1990, the 179th FIS completed its conversion to F-16A/B ADFs. The 179th currently flies F-16s from the Duluth IAP and, on alert, from Tyndall AFB, Florida.

148 TRG/FIG (MN ANG)

RF-4C

64-1023	64-1031	64-1032	64-1033	64-1034	64-1061
64-1066	64-1068	64-1069	64-1074	64-1076[#]	64-1077
64-1079	65-0824	65-0826	65-0830	65-0841	65-0867
65-0891[#]	65-0900	65-0908	65-0942		

F-4D (Received from the 31 TTW, Homestead AFB)

64-0930	64-0952	64-0980	65-0581	65-0585	65-0586
65-0595	65-0608★	65-0629	65-0639	65-0648	65-0652
65-0654	65-0660	65-0687	65-0696	65-0707	65-0739
65-0742	65-0744	65-0748	65-0752	65-0765	65-0767
65-0787	66-0242	66-0270	66-7466	66-7489	66-7490
66-7496	66-7512	66-7635	66-7647		

★ 65-0608 - MiG-21 - September 12, 1972
[#] 64-1076 Destroyed in a mishap on May 25, 1978
[#] 65-0891 Destroyed in a mishap on November 14, 1978

RF-4C 64-1074 is seen here in September 1979. (Maj. General Wayne Gatlin)

F-4D 65-0586 is seen here in February 1984 wearing test tail markings which were not adopted. The test markings were only applied to this aircraft. 0586 was later placed on the Saylor Creek Range, near Mountain Home AFB as a target. (Marty Isham Collection)

RF-4C 65-0900 is seen here on March 27, 1982 with the red MINNESOTA tail stripe. 0900 crashed on March 6, 1994 while assigned to the 117th TRW, Birmingham, Alabama. (Brian C. Rogers)

RF-4C 64-1034 seen on the ramp on March 25, 1977. In standard Southeast Asia Camouflage scheme wears the blue Minnesota tail stripe. 1034 crashed into the Atlantic Ocean following an in flight fire on February 17, 1993 while assigned to the 117th TRW, Birmingham, Alabama. (Brian C. Rogers)

RF-4C 64-1023 seen on the ramp on March 25, 1977. In standard Southeast Asia Camouflage scheme wears the red Minnesota tail stripe. (Brian C. Rogers)

RF-4C 65-0824, seen on the ramp in August 1974 in standard Southeast Asia Camouflage scheme, wears the yellow Minnesota tail stripe. (Don Spering/AIR)

RF-4C 64-1061 is seen on the ramp on July 13, 1982 in standard Southeast Asia Camouflage scheme. The blue tail stripe now includes the name MINNESOTA, and a styled nose black anti-glare paint has been added. 1061 was put on display at the Minnesota ANG Historical Museum. (David F. Brown Collection)

RF-4C 65-0942 seen on the ramp on September 30, 1981 with the name MINNESOTA in the red tail stripe and the styled nose black anti-glare paint added. (Tom Ring)

RF-4C 64-1074 seen in September 1983 with the name MINNESOTA in the yellow tail stripe and the styled nose black anti-glare paint. (David F. Brown Collection)

Minnesota F-4D 65-0595 is seen with California ANG F-4D 65-0747. (Ted Van Geffen)

F-4D MiG killer 65-0608 seen here was credited with the destruction of a North Vietnamese MiG-21 shot down on September 12, 1972 while operating out of Korat RTAFB. The aircraft is painted in the 143rd FIG's standard low visibility markings. 0608 remains on display at Duluth ANG Base. (David F. Brown Collection)

F-4D 65-0639 is seen here on May 5, 1989 in Air Defense Gloss Gray with the standard dark blue Duluth markings. 0639 was used as a target on the Holloman AFB ranges. (Brian C. Rogers)

F-4D 65-0648 is seen here at Ramstein AB, Germany on June 25, 1986 while in Europe for a NATO exercise. (Ted Van Geffen)

F-4D 64-0980 is seen here in September 1986 at the end of the runway waiting its turn to takeoff. 0980 was used as a target on the Cannon AFB, New Mexico range. (David F. Brown)

F-4D 65-0742 is seen here in Hill Grays in June 1987 with the subdued dark gray Duluth markings. (Jerry Geer)

F-4D 66-7635 is seen here in Hill Grays in August 1987 with the subdued dark gray Duluth markings. (Jerry Geer)

153rd TRS, 186th TRG, Key Field, Meridian, Mississippi

The 153rd Tactical Reconnaissance Squadron, 186th Tactical Reconnaissance Group, Mississippi Air National Guard is based at Key Field, Meridian, Mississippi. It flew RF-4Cs for 13 years starting in the fall of 1978 when it received RF-4Cs transferred from RAF Alconbury, United Kingdom. During that period, it made more varied overseas deployments than other ANG tactical units. It deployed to the United Kingdom in September/October 1980 (Coronet Cyro), to Southwest Asia in March 1984 (Sentry Tornado) and to Italy in July/August 1986 (Coronet Lake), The 153rd operated NWDS equipped RF-4Cs prior to conversion to KC-135Rs.

The 153rd TRS traces its origins to the 153rd Observation Squadron, which was organized in September 1939 and which served on active duty in Europe during World War II. As the 153rd Liaison Squadron, the unit distinguished itself during the Battle of the Bulge in the winter of 1944. As the 153rd Fighter Squadron, it flew P-47Ns until reorganized as the 153rd TRS, and converted to RF-51Ds in December 1952. As the 153rd TRS, the squadron was then successively equipped with RF-80As (from June 1955), RF-84Fs (from October 1956, including eleven months on active duty at home during the Berlin Crisis call-up in 1961-1962. In November 1970 the 153rd received RF-101Cs and operated them until replaced with RF-4Cs in September 1978. After receiving its first two KC-135Rs, the 186th Tactical Reconnaissance Group was redesignated as the 186th Air Refueling Group on April 1, 1992.

186 TRG (MS ANG)

RF-4C (Received from the 10 TRW, RAF Alconbury)

65-0931*	65-0935*	65-0942*	65-0945*	66-0400*	66-0401*
66-0402*	66-0408*	66-0413#	66-0415*	66-0418*	66-0421
66-0422	66-0423*	66-0425*	66-0428*	66-0430*	66-0433
66-0449*	66-0478	67-0438	67-0439#	67-0441	67-0442
67-0443	67-0449*	67-0451			

* RF-4C Modified to add AN/ASQ-203(V)1 Navigation and Weapon Delivery System
\# 66-0413 Destroyed in a mishap on June 30, 1985
\# 67-0439 Destroyed in a mishap on March 26, 1980

RF-4C 66-0423 seen on the ramp in July 1979 in standard Southeast Asia Camouflage scheme and white KE (KEy Field) tail code and white serial number configuration. (Jerry Geer)

RF-4C 66-0430 seen on the ramp at 147th on May 17, 1980 in standard Southeast Asia Camouflage scheme and black KE tail code and serial number configuration. (Pete Wilson)

RF-4C 66-0401 seen in August 1981 in Southeast Asia Wraparound Camouflage scheme and the green and gold Mississippi fin flash and black tail code and serial number configuration. (Don Logan Collection)

RF-4C 66-0415 seen on June 11, 1984. (Don Logan Collection)

RF-4C 66-0413 seen on April 4, 1985 in Southeast Asia Wraparound Camouflage scheme and the green and gold Mississippi fin flash. The last two digits of the serial number have been enlarged for identification. (Bob Leavitt)

RF-4C 66-0423 seen on February 2, 1980 in Southeast Asia Wraparound Camouflage scheme and the green and gold Mississippi fin flash. The E in the KE tail code has been temporarily covered in red. (Douglas E. Slowiak/Vortex Photo Graphics)

RF-4C 66-0401 goes vertical. (Don Spering/AIR)

In this view of RF-4C 66-0413 in February 1988, the aircraft is painted in Europe I Camouflage with the enlarged serial number digits. 0413 was written-off on December 3, 1989 by the 117th TRW, Birmingham, Alabama. (Kevin Patrick)

RF-4C 66-0428 seen on the ramp at Luke AFB is painted in Europe I Camouflage scheme. (Kevin Patrick)

RF-4C 66-0402 seen on the ramp on August 30, 1986 is painted in Europe I Camouflage scheme. (Donald S. McGarry)

RF-4C 66-0401 in Europe 1 and 67-0438 in Hill Grays are seen here. (Don Spering/AIR)

RF-4C 66-0425 and his wingman are seen on takeoff during RAM 88. Both aircraft are painted in Hill Grays. (David F. Brown)

RF-4C 66-0449 is seen taxiing in August 1990 painted in Hill Grays. (David F. Brown)

Mississippi ANG RC-4C 65-0931 is seen here on March 9, 1991 on the ramp next to California ANG RF-4C 64-1053. (Paul Minert)

110th TFS, 131st TFW, Lambert Field, St. Louis, Missouri

The 110th TFS was assigned to the 131st TFG, Missouri ANG, at the Lambert-St Louis Airport. The SL tail code reflects their St Louis home. The squadron is located directly across the runway from the McDonnell Douglas factory, the location where all USAF Phantoms had been built. The 110th TFS, 131st TFW had flown Phantoms since the beginning of 1979, when it traded North American F-100D/Fs for F-4Cs, mostly from the 401st TFW at Torrejon AB, Spain. Nicknamed Lindbergh's Own (Charles Lindbergh had been an early member), the 110th TFS began as the pre-World War II 100th Observation Squadron. During World War II it served in the fighter reconnaissance role in the Southwest Pacific.

The 110th was reorganized in 1946 as a Fighter Squadron with F-51Ds, becoming a bombardment squadron in 1952 when it converted to with B-26B/Cs. Returning to its role as a Fighter Squadron in 1957, it has flown F-80Cs, F-84Fs, F-100Cs (which it flew from Toul-Rosieres AB, France, during the Berlin Crisis in 1961-1962) and F-100D/Fs, before receiving F-4Cs in 1979. It first deployed its F-4Cs to the United Kingdom in June/July 1982 for Coronet Cactus. F-4Es mostly from the 57th FWW at Nellis AFB replaced F-4Cs in 1985. The 110th began converting from F-4Es to F-15 A/Bs in May 1991, and achieved F-15 Initial Operational Capability on September 15, 1991. Most of the wing's F-4Es were transferred in 1991 under the Foreign Military Sales program to Turkey and the Republic of Korea.

131 TFW (MO ANG)

F-4C (Received from the 401 TFW, Torrejon AB and other bases)

63-7422	63-7434	63-7530	63-7549	63-7559	63-7588
63-7605	63-7620	64-0661	64-0666	64-0704	64-0747
64-0749	64-0763	64-0772	64-0789	64-0838*	64-0889
64-0902	64-0905	64-0910	64-0911	64-0928	

* 64-0838 - MiG-21 - January 2, 1967

F-4E (Received from the 57 FWW, Nellis AFB and other bases)

67-0353	67-0370	67-0372	67-0376	67-0381	68-0382
67-0391	67-0398	68-0303	68-0311#	68-0337	68-0338**
68-0345	68-0347	68-0348	68-0351	68-0371	68-0374
68-0383	68-0385	68-0403	68-0409	68-0410	68-0448#
68-0462	68-0473	68-0474	68-0476	68-0498	68-0504
68-0528	69-0307	69-7261	69-7267		

** 68-0338 - MiG-21 - May 31, 1972 - MiG-21 - September 16, 1972
# 68-0311 Destroyed in a mishap on April 8, 1990
# 68-0448 Destroyed in a mishap on March 12, 1987

Double MiG killer, F-4E 69-338 seen with F-4 Phantom 30th Anniversary markings. (Boeing)

F-4E two ship in Hill Grays (Boeing)

F-4C 63-7559 in Southeast Asia Wraparound Camouflage seen here in flight 7559 was later put on display at Dobbins AFB, Georgia. (Boeing)

This St. Louis F-4C performs a break away from F-4C 64-0789 revealing the wraparound camouflage pattern on the bottom of the aircraft. (Boeing)

A four ship of St Louis F-4Cs overflys the McDonnell Douglas factory at Lambert Field in St. Louis. (Boeing)

Five St Louis F-4Cs fly past the Gateway Arch on the Mississippi river in St. Louis. (Boeing)

Double MiG killer, F-4E 68-338 is seen on September 17, 1991 on the 131st TFW ramp across the runway from the McDonnell Douglas St. Louis factory which produced all the U.S. Air Force Phantom IIs. (Nate Leong)

68-0338 is seen here in Europe I Camouflage on October 15, 1985. While assigned to the 432nd TRW at Udorn RTAFB in Thailand, 0338 was responsible for the downing of two North Vietnamese MiG-21s, one on May 31, 1972 and the second on September 16 of the same year. (Douglas E. Slowiak/Vortex Photo Graphics)

F-4E 68-0338 is seen in Hill Grays in June 1986 with the SL (St. Louis) tail code of the 131st TFW, Missouri Air National Guard with a non standard serial number presentation. (Mike Grove)

68-0338 is seen here on February 18, 1989 marked as the 131st TFW Commander's aircraft (Douglas E. Slowiak/Vortex Photo Graphics)

During May 1988 68-0338 was marked to celebrate the F-4 Phantom IIs 30th Anniversary. The aircraft remains on display at Lambert Field. (Don Spering/AIR)

F-4C 64-0666 in standard Southeast Asia Camouflage is seen here on the St. Louis ramp in June 1981. 0666 was destroyed as a result of being used for ballistics tests at Aberdeen Proving Grounds, Maryland in November 1986. (Geoff Lebaron)

F-4C 64-0910 in standard Southeast Asia Camouflage with white tail markings is seen here on the St. Louis ramp on July 15, 1980. 0910 was later used as a target on the Nellis AFB Range Complex. (Douglas E. Slowiak/Vortex Photo Graphics)

F-4C 64-0928 is seen here on the St. Louis ramp in August 1980. It was later used as a target on the Eglin Range Complex. (Douglas E. Slowiak/Vortex Photo Graphics)

F-4C 64-0749 in Southeast Asia Wraparound Camouflage with black tail markings including a SL tail code is seen here in September 1983. 0749 was used as a battle damage repair aircraft at Homestead AFB, Florida. (David F. Brown Collection)

F-4C 64-0789 in Europe I and black tail markings is seen here in August 1984 on the St. Louis ramp. (David F. Brown)

F-4E 69-7267 is seen in Europe I Camouflage and the SL (St. Louis) tail code of the 131st TFW, Missouri Air National Guard in June 1985. 131st F-4Es wore a shark mouth similar to the one used on 388th TFW F-4Es in Southeast Asia. 7267 was converted to an F-4G. (David F. Brown)

F-4E 68-0311 is seen in Europe I Camouflage and the 131st shark mouth on April 20, 1987. 0311 was "written off" by the 131st TFW on April 8, 1990. (Pete Wilson)

F-4E 68-0403 is seen in Hill Grays taxiing out for takeoff at McConnell AFB in April 1990. 0403 was transferred to the Turkish Air Force under the Peace Diamond program. (Don Logan)

F-4E 68-0351 prepares for engine start prior to departure from Luke AFB in June 1987. 0351 was transferred to the Turkish Air Force under the Peace Diamond program. (Douglas E. Slowiak/Vortex Photo Graphics)

F-4E 68-0504 is seen here in May 1988 at St. Louis is painted in Hill Grays Camouflage Scheme. Visible in the left forward Sparrow well is an AN/AVQ-23A/B Pave Spike LASER designator pod. (David F. Brown Collection)

173rd TRS, 155th TRG, Lincoln Municipal Airport, Lincoln, Nebraska

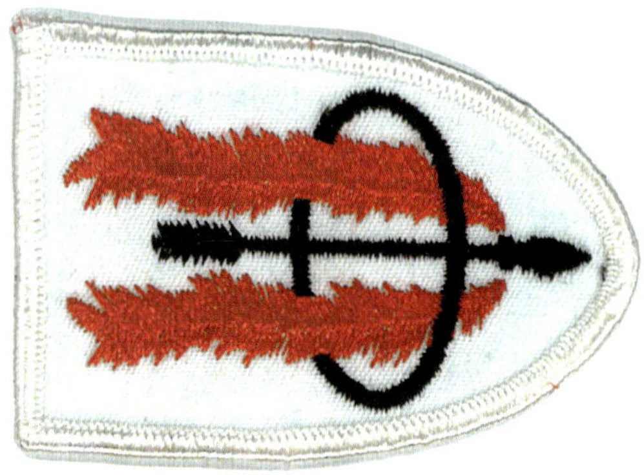

The 173rd Tactical Reconnaissance Squadron, 155th Tactical Reconnaissance Group, Nebraska Air National Guard was based at Lincoln Municipal Airport, Lincoln, Nebraska. The 173rd operated RF-4Cs for over 22 years. The Nebraska unit was organized in July 1946 as the 173rd Fighter Squadron (SE) and was equipped with F-51Ds. In 1948, it became one of the first Guard units to fly jets when it received F-80C Shooting Stars. However, these jets were returned to Active Duty USAF units during the Korean Conflict. Remaining in the U.S. during the Korean Conflict, The 173rd was called to active duty and returned to F-51Ds. After return to state control, the 173rd briefly operated as a fighter-bomber squadron with F-51Ds before converting again to F-80Cs and becoming the 173rd FIS. The Shooting Stars were replaced by F-86Ds in January 1957 and F-86Ls in late 1959.

In May 1964 the 173rd began it mission as a reconnaissance squadron when it converted to RF-84Fs and was redesignated the 173rd TRS. In February 1972 it converted to RF-4Cs. It first deployed to Turkey in May 1980 for Coronet Cannon. The 173rd RF-4Cs were also equipped with the Boeing NWDS. Tactical was dropped from the unit designation and, effective March 16, 1992, it became the 173rd RS. The 155th Tactical Reconnaissance Group received KC-135Rs and was redesignated 155th Air Refueling Group on January 1, 1994.

155 TRG (NE ANG)

RF-4C

64-0998 64-0999 64-1023* 64-1039* 64-1050* 64-1062
64-1063* 64-1066 64-1079* 65-0824* 65-0828* 65-0830
65-0831 65-0838* 65-0840 65-0853* 66-0859* 65-0668
65-0874* 65-0878 65-0902* 65-0903* 65-0907* 65-0911*
65-0917* 65-0920 65-0923* 65-0925 65-0928* 65-0931*
65-0932* 65-0934# 65-0935* 65-0939* 66-0417* 66-0418*
66-0428* 66-0449*

* RF-4C Modified to add AN/ASQ-203(V)1 Navigation and Weapon Delivery System
\# 65-0934 Destroyed in a mishap on November 20, 1977

Nebraska Tails (George Cockle)

RF-4C 64-1066 is seen here in June 1990 painted as the 155 TRG Commander's aircraft. (David F. Brown)

RF-4C 65-0878 is seen here in August 1989 with special NEBRASKA HUSKERS tail markings. (Jerry Geer)

RF-4C 65-0824 is seen here in August 1987 at Lincoln MAP Nebraska in a special version of Europe I Camouflage. (James Goodall)

RF-4C 66-0417 is seen here in July 1980 carrying CIHANDA SULH on the nose. Added to the aircraft as a result of a deployment to Turkey, CIHANDA SULH is a greeting in Turkish. (George Cockle)

RF-4C 64-1062 is seen here on September 10, 1988 with special rudder paint indicating the aircraft had taken part in ALR-74 ECM system tests. (Brian C. Rogers)

RF-4C 65-0838 in standard Southeast Asia Camouflage with white tail markings is seen here on February 7, 1976. (D. J. Fisher)

RF-4C 65-0828 in standard Southeast Asia Camouflage with special yellow and blue stripes on the fuselage and a cartoon of a bulldog on the intake is seen on November 18, 1981. (Don Logan Collection)

RF-4C 65-0911 in standard Southeast Asia Camouflage with white tail markings is seen here on March 1, 1978. (Ben Knowles)

RF-4C 64-0999 in standard Southeast Asia Camouflage with white tail markings and a green and gold NEBRASKA tail stripe is seen here on May 31, 1979. (Marty Isham Collection)

In this view of RF-4C 64-0999 in standard Southeast Asia Camouflage, the aircraft has black tail markings with the green and gold NEBRASKA tail stripe.(George Cockle)

RF-4C 65-0939 in standard Southeast Asia Wraparound Camouflage with black tail markings and a green and gold NEBRASKA tail stripe and LINCOLN rudder is seen here in July 1982. In addition a modified "SPOOK", the Phantom II mascot, has been added to the intake splitter. (Mike Grove)

RF-4C 64-1066 in standard Southeast Asia Wraparound Camouflage is seen here in November 1982. In addition to the green and gold NEBRASKA tail stripe, a green and gold rudder is painted with LINCOLN and a star added rudder. (George Cockle)

RF-4C 65-0853 in Europe I is seen taxiing here on November 1986 as part of RAM 86.The open doors of the photo flash system can be seen on the tail. (George Cockle)

RF-4C 65-0838 in Europe I with black tail markings is seen here in September 1988. (David F. Brown)

RF-4C 65-0903 in Hill Grays Camouflage and yellow tail stripe is seen landing in August .1990. (David F. Brown)

RF-4C 65-0902 in Hill Grays Camouflage and blue tail stripe is seen on August 28, 1992. (Norris Graser)

RF-4C 65-0917 in Hill Grays Camouflage and yellow tail stripe is seen in November 1989. (Jerry Geer)

RF-4C 65-0939 in Hill Grays Camouflage and red tail stripe is seen taxiing back following a landing on September 13, 1991. (David F. Brown)

RF-4C 64-0999 is seen here in Hill Grays Camouflage and blue tail stripe. (David F. Brown Collection)

RF-4C 65-0920 with a new style rudder marking carrying LINCOLN and a star is seen on July 18, 1987. (Doug Remington)

RF-4C 65-0830 in Hill Grays Camouflage is seen on June 4, 1989. (David F. Brown)

DI JOE Nose art photographed in June 1993 was painted on 65-0838. (Don Logan)

HAWG WILD Nose art on 65-0824 was photographed on April 1, 1990 (Don Logan)

WALLY GATOR Nose art was painted on 65-0917. (Don Logan Collection)

SCRAPPY DOO Nose art was painted on 65-0907 (Don Logan)

DEFENDER OF FREEDOM Nose art on 65-0931 was photographed on September 3, 1998 (Don Logan)

GIRLS JUST WANT TO HAVE FUN Nose art was painted on 66-0418. (Don Logan)

192nd TRS, 152nd TRG, Reno-Cannon International Airport, Reno, Nevada

The High Rollers were activated as the 192nd Fighter Squadron G (SE) in April 1948, equipped with F-51 Ds. They served on active duty in the U.S. during 1951 and 1952, returning to state control in December 1952. The unit, redesignated the 192nd Fighter-Bomber Squadron, moved from Stead ANG Base to the Reno Municipal Airport in January 1953. In 1954 the unit was redesignated the 192nd FIS and began jet training, completing its conversion from F-51Ds to F-86As in November 1955. In August 1958 the day-fighter F-86As were replaced by all-weather F-86Ls. After converting to RB-57Bs, the squadron was redesignated the 192nd TRS in April 1961. In 1965 the 162nd converted to RF-101Hs. The unit served on active duty in the U.S. and Japan during the 1968-69 Pueblo Crisis call-up, and, in 1971, received RF-101Bs, a reconnaissance version of the F-101B Voodoo interceptor.

RF-4C Phantoms were first received during the summer of 1975 when the 192nd TRS converted from its unique McDonnell RF-101Bs. The 192nd was the first unit to operate RF-4Cs fitted with the Boeing AN/ASQ-203(V)1 Navigation and Weapon Delivery System (NWDS). The 192nd was called to active duty for Desert Shield/Desert Storm and flew 412 sorties before being released from active duty on April 20, 1991.

In 1991, after returning from flying combat operations from Sheikh Isa AB, Bahrain, during Operation Desert Storm, the Nevada ANG flying unit began preparations for its conversion from RF-4Cs to F-4Gs. Although the unit did receive F-4Gs, a single F-4G Wild Weasel (69-7580) was delivered to Reno and painted it in the distinctive High Rollers markings. Plans for the conversion to F-4Gs were canceled and the 192nd RS (a designation that became effective on March 16, 1992) continued to fly RF-4Cs until converting to C-130s in fall 1995.

This tail shot of High Rollers RF-4C 64-1068 was taken on August 8, 1991. (Phillip Huston)

152 TRG (NV ANG)

RF-4C

64-1003#	64-1005*	64-1006*	64-1014*	64-1016*	64-1017*
64-1019	64-1020#	64-1021*	64-1022*	64-1029*	64-1030*
64-1039	64-1041*	64-1050*	64-1063*	64-1065*	64-1068*
64-1079*	65-0823*	65-0832*	65-0864*	65-0870*	65-0874*
65-0875*	65-0876*	65-0881*	65-0886*	65-0890*	65-0892*
65-0897*	65-0901*	65-0902*	65-0908*	65-0910*	65-0911*
65-0932*	66-0408*	66-0417*	66-0475	67-0456*	68-0593*

* RF-4C Modified to add AN/ASQ-203(V)1 Navigation and Weapon Delivery System
\# 64-1003 Destroyed in a mishap on November 21, 1980
\# 64-1020 Destroyed in a mishap on April 27, 1988

RF-4C 65-0870, of the HIGH ROLLERS, Nevada Air National Guard taxis out for takeoff during August 1990. (David F. Brown)

While operating in Saudi Arabia following Desert Storm, as part of Operation Southern Watch, the 152nd TRG shared aircraft with the 117th TRW from Birmingham, Alabama. This Alabama jet, 64-1065 photographed in April 1994, has had its 117th TRW tail stripe replaced with a High Rollers Tail cap. (Ben Knowles)

F-4G 69-7580, the only F-4G Wild Weasel assigned to the 152nd TRG, is seen here on August 8, 1991 in the unit's Reno, Nevada hangar during a maintenance inspection. Plans to convert from the RF-4C to the F-4G Wild Weasel were canceled. 69-7580 was converted to QF-4G AF-182. (Phillip Huston)

RF-4C 64-1016, the High Rollers first RF-4, Is seen here at Reno, Nevada in May 1985. It was later converted to an AN/ASQ-203 NWDS equipped jet. (Don Logan)

RF-4C 64-1005 is seen here at Reno in September 1981 wearing the early HIGH ROLLERS markings consisting of a white tail stripe with five blue stars. (Don Logan Collection)

RF-4C 64-1006 is seen here in partial HIGH ROLLERS markings. Red and blue was later added to the white chevron on the tail and RENO added to the white rudder. 1006 was later transferred to Spain as CR12-53. (Jim Rotramel)

RF-4C 64-1022 is seen here with the white tail stripe with blue stars in February 1980. A white rudder with RENO has been added to the High Rollers markings. 1022 was later put on display at Susanville Municipal Airport, Susanville, California. (David F. Brown Collection)

The markings on this Europe I painted RF-4C (64-1021) have changed again, this time for Reconnaissance Air Meet (RAM) 86. The white tail stripe now contains HIGH ROLLERS, and the RENO on the rudder has been replaced with NEVADA. (George Cockle)

This two-ship of Europe I painted RF-4Cs (64-1005 and 64-1021) taxi out for takeoff in November 1986 during RAM 86. (George Cockle)

RF-4C 64-1068 is seen here in September 1985 with the HIGH ROLLERS tail stripe and a plain rudder. (Douglas E. Slowiak/Vortex Photo Graphics)

RF-4C 65-0890 is seen here on August 25, 1984. (Kevin Patrick)

RF-4C 64-1068 is seen here painted in Hill Grays with the Reconnaissance Air Meet (RAM) 82 on the intake and BATTLE BORN NEVADA emblem on the rudder. (Jerry Geer)

RF-4C 65-0897 is seen here in October 1990 with the RAM 82 markings and a large red 1 on the intake splitter. (Ben Knowles)

RF-4C 64-1021 is seen here in August 1988 with RAM 82 markings. (David F. Brown)

RF-4C 65-0897 is seen here on landing during August 1990 just as the pilot chute for landing drag chute is deploying. A NWDS jet, 0897 was later transferred to Spain as CR12-58. (David F. Brown)

RF-4C 65-0911 is seen here in Hill Grays in August 1995 with the HIGH ROLLERS tail stripe. (Jerry Geer)

RF-4C 65-0932 is seen here on July 15, 1995 in Hill Grays with the HIGH ROLLERS tail stripe. (Keith Snyder)

141st TFS, 108th TFW, McGuire AFB, New Jersey

The 141st TFS, 108th TFW, New Jersey Air National Guard was based at McGuire Air Force Base, near Trenton, New Jersey. The 141st was activated as a National Guard unit in May 1949, as the 141st Fighter Squadron (SE) and flew F-47Ds for three and one half years, including 21 months on active duty in the United States during the Korean conflict. The 141st Fighter-Bomber Squadron was re-equipped with F-51Hs in December 1952 and with F-86As in February 1954. The being redesignated 141st FIS in July 1955, the squadron received F-86Es in February 1956. In the spring of 1958, the 141st received F-84Fs Thunderstreaks, and was once again redesignated, this time in November 1958 as 141st TFS in November 1958. The unit was called to active duty in October 1961 and was based at Chaumont AB, France, until October 1962. After its return from France, the 141st was equipped with F 86Hs in October 1962 and F-105Bs in May 1964.

The 141st converted to F-4Ds from the 56th TFTW at MacDill AFB in spring 1981. An NJ tail code was applied to the aircraft. F-4Es from Ramstein AB Germany replaced the F-4Ds in July 1985. After flying Phantom IIs from McGuire AFB since the spring of 1981, the 141st converted to KC-135Es in the fall of 1991 and was redesignated the 141st AREFS, 108th AREFW, on October 19, 1991. Many of the wing's F-4Es were transferred under FMS to Korea and Turkey.

While equipped with F-4Ds, the 141st TFS first deployed to Norway in September 1982 (Coronet Rawhide) and, as a fighter unit, the 141st made its last deployment in May 1991 when it sent F-4Es to Ecuador for Blue Horizon.

108 TFW (NJ ANG)

F-4D (Received from MacDill AFB)

65-0631	65-0644	65-0647	65-0660	65-0667	65-0674#
65-0680	65-0696	65-0701	65-0703	65-0712	65-0739
65-0741	65-0742	65-0756	65-0775	66-0259	66-0262
66-0270	66-0279	66-7478	66-7496	66-7498	66-7512

65-0674 Destroyed in a mishap on July 15, 1983

F-4E (Received from Ramstein AB)

67-0343	67-0360	67-0395	68-0326	68-0346	68-0357
68-0358	68-0370	68-0375	68-0377	68-0378	68-0379
68-0386	68-0387	68-0388	68-0390	68-0392	68-0401
68-0413	68-0459	68-0460	680464	68-0494	68-0508#
68-0514	68-0526	68-0527	68-0529	68-0533	68-0534

68-0508 Destroyed in a mishap on June 18, 1986

F-4D 65-0667 is seen here loaded with a centerline gun and BDU-33 practice bombs. (Don Spering/AIR)

Opposite: F-4E 67-0343 goes vertical. (Don Spering/AIR)

NJ
343

From top to bottom F-4Es 68-0527, 68-0526, and 68-0378 are examples of the three F-4 camouflages. (Don Spering/AIR)

F-4Es in each of the three camouflages, Southeast Asia, Hill Grays, and Europe I are seen here in front of the 108th TFW hangar on October 19, 1987. (David F. Brown)

F-4E 68-0370 is seen here over the snow capped mountains of western New Jersey. (Don Spering/AIR)

The TIGERS of the 141st TFS applied tiger heads to their F-4Es. This full color head has been applied to a aircraft painted in Europe I Camouflage. (David F. Brown)

As seen here on December 6, 1987, each of the different camouflage schemes used a different color for the tiger head. (David F. Brown)

F-4E 68-0527 is seen here in Hill Grays painted as the 108th TFW Commander's aircraft. The NJ tail code is derived from New Jersey. (Ben Knowles)

F-4E 68-0346 is seen here in Hill Grays painted as the 141st TFS Commander's aircraft. 0346 was transferred to Turkey under the Peace Diamond program in August 1991. (David F. Brown)

New Jersey F-4Ds lined-up on the ramp. (Don Spering/AIR)

Three New Jersey F-4Ds, seen here on November 1983, line-up on the runway waiting for the fourth member of their flight. (David F. Brown)

F-4D 65-0701 taxis past the VORTAC station at McGuire AFB, New Jersey (David F. Brown)

F-4D 65-0644 is seen here in Southeast Asia Wraparound Camouflage scheme on May 15, 1984. (David F. Brown)

F-4D 65-0667 is seen here in Europe I Camouflage scheme in October 1984. (Don Logan Collection)

F-4D 65-0703 is seen here in Europe I Camouflage scheme taxiing for takeoff during July 1984. (David F. Brown)

F-4D 65-0701 is seen here in Europe I Camouflage scheme in August 1985. (David F. Brown Collection)

F-4E 67-0343 is seen here in Southeast Asia Wraparound Camouflage at McGuire AFB on July 17, 1982. (Barry Roop)

F-4E 68-0459 is seen in March 1987 as it taxis past the VORTAC station at McGuire AFB, New Jersey. (David F. Brown)

F-4E 67-0343 is seen here in Southeast Asia Wraparound Camouflage with a tiger head added on December 6, 1987. 0343 was converted to QF-4E AF-111 in May 1996. (David F. Brown)

F-4E 67-0377 is seen here in Europe I at McGuire AFB in March 1986. (David F. Brown Collection)

F-4E 67-0377 is seen here loaded with practice AIM-9 sidewinder missiles on April 18, 1986. 0377 was transferred to the Greek Air Force in November 1991. (David F. Brown Collection)

F-4E 68-0413 is seen here in Europe I at McGuire AFB on March 21, 1990. (Barry Roop)

F-4E 68-0378 is seen here on June 10, 1990 parked on the ramp in front of the 108th CAMS hangar at McGuire AFB painted in Europe I with a full color tiger head. 0378 was transferred to the Republic of Korea Air Force in August 1990. (David F. Brown)

F-4E 68-0386 is seen here on July 27, 1987 painted in Europe I with the basic one color tiger head. 0386 was transferred to the Republic of Korea Air Force in August 1990. (David F. Brown)

Another Europe I F-4E (68-0413) is seen here on December 6, 1987 painted with the basic one color tiger head. 0413 was transferred to the Republic of Korea Air Force in July 1991. (David F. Brown)

F-4E 68-0370 is seen here over ice covered mountains (Don Spering/AIR)

Two F-4Es painted in Hill Grays painted F-4Es, 108th TFW Commander's aircraft 68-0527 and 68-0357 are seen November 16, 1990 parked awaiting aircrews. (Barry Roop)

F-4E 68-0377 is seen here in December 1990 on landing just as the pilot chute is beginning to deploy the main drag chute. (David F. Brown)

F-4E 68-0533 is seen here painted in Hill Grays in July 1988. 0533 was transferred to the Republic of Korea Air Force in September 1991. (David F. Brown Collection)

136th FIS, 107th FIG, Niagara Falls International Airport, New York

The 136th Fighter Interceptor Squadron, 107th Fighter Interceptor Group, New York Air National Guard is based at Niagara Falls International Airport, Niagara Falls, New York. The 136th Fighter Squadron (SE) was activated at Niagara Falls on December 1948 and, still equipped with its original F-51Ds, flew on active duty from its home base during the Korean War. As the 136th FIS, it was re-equipped with F-51Hs in late 1952 and with F-94Bs in January 1954. Redesignated 136th TFS in October 1957, the unit flew F-100Cs for fourteen years, including while on active duty at its home base in 1961-1962 and deployed at Tuy Hoa AB, South Vietnam, in 1968-1969. In June 1971, the 136th once again became a fighter interceptor unit. As an FIS it has flown F-101 B/Fs, F-4Cs (from spring 1982) and F-4Ds (from fall 1986). The 136th received its F-4Cs from the 58th TFW at Luke AFB and a few from the 170th TFS, Illinois ANG. They received F-4Ds from 31st TFW Homestead AFB and 184th TFG Kansas ANG. While equipped with Phantoms, the 136th FIS maintained an alert detachment at Charleston AFB. In 1990, the 136th began its conversion to F-16A/B in ADF configuration. The 136th Fighter Squadron received KC-135Rs and was redesignated as the 136th Air Refueling Squadron on July 1, 1994.

107 FIG (NY ANG)

F-4C (Received from Luke AFB)

63-7414	63-7428	63-7433	63-7465	63-7506	63-7523
63-7530	63-7541*	63-7557	63-7566	63-7568	63-7581
63-7598	63-7633	63-7662	64-0659	64-0660***	64-0661
64-0666	64-0691	64-0750	64-0763	64-0822	64-0836
64-0860	64-0891	64-0899			

* 63-7541 - MiG-21 - November 5, 1966

*** 64-0660 - MiG-17 - May 12, 1966 - MiG-17 - May 14, 1967 - MiG-17 - June 5, 1967

F-4D (Received from Homestead AFB and Kansas ANG)

64-0965	65-0692	65-0597	65-0609	65-0626	65-0631
65-0638	65-0647	65-0659	65-0677	65-0681	65-0684
65-0692	65-0699	65-0738	65-0752	65-0753	65-0763
65-0777	65-0781	65-0789	65-0792	66-0259	66-0277
66-7456	66-7469	66-7472	66-7485	66-7487	66-7488
66-7491	66-7497				

F-4C 63-7485 is seen here on and active air scramble from the Charleston AFB Detachment in February 1989. (Doug Barbier)

Right: F-4Cs and Ds during their career in the AFRES/ANG picked up the nickname Rhino. As a result, as seen here, the 136th FIS painted Rhino head on the nose of their aircraft. (Don Logan)

Below: F-4C 64-0660 seen here in June 1983 was responsible for the downing of three North Vietnamese MiG-17 fighters. The first MiG-17 was shot down on May 12, 1966 while 0660 was assigned to the 35th TFW. The second and third MiG-17 were shot down while 0660 was assigned to the 366th TFW, on May 14, 1967 and June 5, 1967. (David F. Brown)

Triple MiG Killer F-4C 64-0660 seen here in January 1985 remained with the 136th FIS and was eventually placed on display at Niagara Falls International Airport, New York. (David F. Brown)

F-4C 63-7541 seen here on November 25, 1986 was responsible for the downing a North Vietnamese MiG-21 fighter on November 5, 1966 while assigned to the 366th TFW. 7541 was later placed on the Nellis AFB, Nevada ranges and used as a target. (Don Logan Collection)

The 136th FIS adopted a rainbow painted rudder as part of their unit markings. F-4C 64-0659 is seen here in Air Defense gray, painted with the rainbow rudder in June 1984. (David F. Brown)

F-4C 63-7557 is seen here on August 1984 in Southeast Asia Wraparound Camouflage, painted with the rainbow rudder. (David F. Brown)

F-4C 63-7581 is seen here on September 23, 1984 painted with the rainbow rudder. (David F. Brown Collection)

F-4C 66-0277 is seen here on June 8, 1988 in Europe 1 camouflage, before the rudder was painted. (David F. Brown)

F-4C 64-0860 is seen here on November 25, 1986 marked as "The Spirit of North Charleston". The 136th FIS pulled NORAD alert at Charleston AFB, South Carolina. (Tom Brewer Collection)

F-4C 64-0659 is seen here in Air Defense gray, painted with the rainbow rudder on May 11, 1985. 0659 was retired to Aberdeen Proving Grounds, Maryland where it was used as a ballistics test aircraft. (David F. Brown Collection)

F-4C 63-7568 is seen here in June 1983 in Air Defense gray, painted with the rainbow rudder and TAC style serial number presentation. 7541 was later placed on the Nellis AFB, Nevada ranges and used as a target. (David F. Brown)

F-4C 63-7428 is seen here in Hill Grays on November 1, 1986 painted with the rainbow rudder and an unusual single line serial number presentation. (Don Logan Collection)

F-4D 65-0692 is seen here on August 3, 1988 in Air Defense gray painted with a new 136th tail insignia. (Don Logan Collection)

F-4D 66-7485 is seen here on October 11, 1988 in Hill Grays painted with a rhino nose and the new 136th tail insignia. (Don Logan Collection)

F-4D 66-0259 is seen here in November 1990 in Hill Grays with the rhino nose and the new 136th tail insignia. 0259 was later put on display at McGuire AFB, New Jersey. (Don Logan Collection)

F-4D 66-7485 is seen here on June 6, 1988 in Hill Grays painted with the new 136th tail insignia in blue and light green. (David F. Brown)

F-4D 65-0692 is seen here on June 7, 1987 in Air Defense gray with a plain tail. (David F. Brown)

F-4D 65-0753 is seen here on July 18, 1987 in Hill Grays with two "Kill" markings on the intake splitter. (Doug Remington)

178th FIS, 119th FIG, Hector International Airport, Fargo, North Dakota

The Happy Hooligans of the North Dakota ANG are based at the Hector Field in Fargo. They flew F-4Ds between March 1977 and April 1990, prior to converting to F-16A ADFs. The 178th with its F-4Ds, deployed twice overseas, flying 'Bear' patrols in Iceland in 1984 and becoming the first core unit to provide personnel and equipment for Operation Creek Klaxon at Ramstein AB, Germany, in 1986-1987. In January 1947, the 178th Fighter Squadron (SE) flew North American F-51Ds/F-51Ds until June 1954 (including active duty in Georgia and California during the Korean War), Lockheed F-94A/Bs and then F-94Cs until June 1958, Northrop F-89Ds and then F-89Js until July 1966, Convair F-102A/TF-102As until November 1969 and McDonnell F-101Bs until converting to Phantom IIs in March 1977.

The 178th FIS received their F-4D from Hill AFB, George AFB, Holloman AFB and RAF Lakenheath. The Hooligans presently fly F-16A/B ADF Falcons.

119 FIG (ND ANG)

F-4D (Received from Various Units)

64-0937 – 02	64-0938 – 03	64-0939 – 04	64-0942 – 05	64-0945 – 06	64-0949 – 07
64-0952 – 11	64-0953 – 12	64-0956 – 13	64-0959 – 14	64-0963 – 15	64-0965 – 16
64-0968 – 17	64-0970 – 21	64-0972 – 22	64-0973 – 23	64-0975 – 24	64-0976 – 25
64-0977 – 26	64-0979 – 27	65-0629	65-0647	65-0696	65-0760
65-0769	66-7478	66-7498			

The 2-digit numbers following the serial numbers are NORAD tactical numbers that were painted in black outline lettering on the rudders of the F-4Ds. Some of the Southeast Asia camouflaged aircraft had the last two digits of the serial numbers painted in a larger size, these were not NORAD identification numbers, but rather used to make identification of the aircraft by 178th FIS personnel easier.

HAPPY HOOLIGANS F-4D 64-0939 is seen here on May 31, 1988 over the North Dakota countryside. (Brian C. Rogers)

Air Defense Gray F-4D 64-0963 sits on the Hector Field ramp in August 1989 next to Hill Grays F-4D 64-0937. (Don Logan)

During 1989, F-4D 66-7498 seen here in August 1989 was specially painted in honor of the North Dakota Centennial. (Don Logan)

As part of a test modification, the 178th FIS added an infrared detector to the nose of F-4D 64-0945. The detector was similar to the ones used by the F-102. (Douglas E. Slowiak/Vortex Photo Graphics)

Happy Hooligans F-4D 64-0937 is seen here in November 1977 in standard Southeast Asia Camouflage scheme. The serial number presentation on the tail has been modified with the last two digits "37" being used as the NORAD identification number. 0937 was retired to Holloman AFB, New Mexico to be used as a target on the range complex. (Don Logan)

Happy Hooligans F-4D 64-0945 is seen here on an icy North Dakota ramp in January 19, 1980. 0945 was retired to Holloman AFB, New Mexico to be used as a target on the range complex. (Douglas E. Slowiak/Vortex Photo Graphics)

F-4D 66-7498 is seen here in Europe I Camouflage on January 10, 1986. (Kevin Patrick)

F 4D 66 7498 is seen here at Luke AFB on January 10, 1986. (Kevin Patrick)

F-4D 64-0963 is seen here as it passes a VORTAC station in June 1989. (David F. Brown)

F-4D 64-0976, nicknamed "ABAGAIL" is seen here in May 1989. The radome and nose are painted in a non-standard gray. In December 1997 0976 was retired to Holloman AFB, New Mexico to be used as a target on the range complex. (David F. Brown)

F-4D 64-0953 is seen here on June 21, 1986 with its hook down. (Kevin Patrick)

F-4D 64-0952 is seen returning from a mission in October 1982, trailing its red, white, and blue drag chute. Both crewmembers are wearing the orange flight suits commonly used by air defense aircrews. 0952 was surplused to Avtel Flight Test at Mojave airport, California where it flew as N401AV. (Don Logan Collection)

F-4D 64-0939 is seen here on the ramp in May 1984. (Charles T. Robbins)

F-4D 64-0949 is seen taxiing out for a mission during August 1989. The 178th FIS flew some of their F-4Ds with modified radomes, removing the lower IR Sensor fairing which was no longer used. 0949 was retired to Holloman AFB, New Mexico and used as a range target. (Don Logan)

F-4D 65-0647 is seen here in August 1989 with a full air-to-air weapons load; AIM-7 Sparrow missiles in the fuselage missile wells, AIM-9 Sidewinders on the inboard pylons, and a 20mm gun pod on the centerline station. 0647 was retired to Cannon AFB, New Mexico and used as a range target. (Don Logan)

F-4D 66-7498 is seen here on the ramp in October 1989. (Jerry Geer)

114th TFTS, 142nd FIG, Kingsley Field, Klamath Falls, Oregon

The 114th TFS, based at Kingsley Field near Klamath Falls, was organized on February 1, 1984 to serve as an Replacement Training Unit (RTU) for F-4 crews assigned to Fighter Interceptor Squadrons (FIS) of the Air National Guard. This numerical designation had previously been used by a unit of the New York ANG, organized in June 1947 as the 114th Bombardment Squadron (Light) flying Douglas B-26B/C Invaders from Floyd Bennett Field. The New York ANG squadron was redesignated 114th FIS in June 1957 flying Lockheed F-94B Starfires when it was inactivated in September 1958.

In 1982 the Air Force announced a proposal to establish an air defense schoolhouse for F-4 Phantom II aircrews under the Air National Guard. In mid-November 1988, the last F-4 class graduated. As a unit of the Oregon ANG, the 114th TFTS flew F-4Cs from its activation in 1984 until its conversion to F-16 ADFs during spring 1989. Beginning in November, 1997, the Wing began plans to change aircraft from the F-16 to the F-15 Eagle. The unit remained as a schoolhouse helping to relieve shortages in F-15 training quotas for the total force. The last F-16s were flown out on February 26, 1998 to Tucson, Arizona. After extensive maintenance and pilot training, F-15 flying operations began in June 1998.

114 TFTS (OR ANG)

F-4C

63-7412	63-7428	63-7448	63-7460	63-7479	63-7482
63-7490	63-7495	63-7529	63-7532	63-7537	63-7549
63-7555	63-7581	63-7584	63-7620	63-7622	63-7633
63-7662	63-7704★	64-0655	64-0673★	64-0754	64-0775
64-0804	64-0828	64-0838★	64-0888	64-0890	64-0899
64-0905	64-0911				

★ 63-7704 - MiG-17 - May 14, 1967
★ 64-0673 - MiG-17 - May 20, 1967
★ 64-0838 - MiG-21 - January 2, 1967

F-4C 64-0838, on January 6, 1967 while assigned to the 8th TFW at Ubon RTAFB, was credited with shooting down a North Vietnamese MiG-21. (Douglas E. Slowiak/Vortex Photo Graphics)

F-4C 63-7507 is seen here in July 1985 in Southeast Asia Wraparound Camouflage scheme. The only unit markings applied at this time is OREGON at the top of the vertical stabilizer. 7507 was later used as a GF-4C ground trainer at Sheppard AFB, Texas. (David F. Brown)

F-4C 64-0899, BEAUTY JET, is seen here in February 1988. 0899 was retired to Holloman AFB, New Mexico and used as a range target. (Don Logan Collection)

MiG killer F-4C 64-0838 is seen taxiing out for takeoff in September 1986. (David F. Brown)

F-4C 63-7482 is seen here on August 22, 1986. The tail has the chevrons left over from the unit markings it wore while assigned to the 171st FIS at Selfridge AFB, Michigan. It was retired and put on display at Minnesota ANG Historical Museum. (Doug Remington)

F-4C 63-7537 is seen here at Carswell AFB, Texas on January 30, 1985. 7537 was retired to Holloman AFB, New Mexico and put on display. (Brian C. Rogers)

F-4C 64-0775 is seen here at Luke AFB, Arizona in December 1986. (Kevin Patrick)

F-4C 63-7584 is seen here in front of the 114th TFTS Hangar at Klamath Falls, Oregon in August 1985. 7584 was put on display at McChord AFB, Washington in July 1988. (David F. Brown)

F-4C MiG Killer 64-0838 with its radome is seen on August 8, 1988 undergoing maintenance on the ramp at Klamath Falls. 0838 was put on display at The US Space and Rocket Center, Huntsville, Alabama in June 1989. (Phillip Huston)

F-4C 63-7620 is seen here on January 24, 1987 in Europe I Camouflage scheme and full 114th TFTS unit markings. 7620 was later used as a range target on the Nellis AFB, Nevada range complex. (Bob Greby)

F-4C 63-7555 is seen here on October 6, 1987 in Hill Grays scheme. The aircraft is carrying the 114th TFTS Commander's travel pod on the left inboard under wing pylon. 7555 was put on display at the Yankee Air Museum, Michigan. (Brian C. Rogers)

F-4C 63-7532 is seen here on October 1, 1986 in Hill Grays paint scheme and full 114th TFTS unit markings. 0838 was put on display at Barksdale AFB, Louisiana. (Doug Remington)

F-4C 63-7633 is seen here on August 9, 1988 in Hill Grays paint scheme and a non-standard Air Defense style serial number presentation. (Phillip Huston)

F-4C 63-7633 is seen here taxiing back from a mission in July 1987. 7633 was later used as a range target on the Saylor Creek Range complex south of Mountain Home AFB, Idaho. (David F. Brown)

123rd FIS, 142nd FIG, Portland International Airport, Oregon

Oregon ANG's 123rd FIS Red Hawks was part of the 142nd FIG and was based at the Portland International Airport, Oregon Between fall 1982 and fall 1989, the 123rd FIS flew F-4Cs The 123rd FIS was originally activated as the 123rd Observation Squadron, was organized in April, 1941. Five months later, the Oregon unit began flying ASW missions and served as a training unit in the United States. It was redesignated the 35th Photographic Reconnaissance Squadron in August 1943 and equipped with Lockheed F-5s (Reconnaissance version of the P-38).

Redesignated the 123rd Fighter Squadron (SE), the unit was initially equipped with F-51Ds and then flew F-86Fs while on active duty at Portland AFB during the Korean call-up. After returning to state control in December 1952, the 123rd FIS converted to F-86As. It received F-94Bs in October 1955, F-89Ds in June 1957, F-89Hs In November of the same year, F-89Js in September 1960, F-102A/TF-102As in January 1966 and F-101B/Fs in March 1971. It converted to F-4Cs in 1982 receiving jets from the 170th TFS, Illinois ANG. The 123rd FIS phased out the ANG's last F-4Cs and T-33As in the fall of 1989. The unit received its first F-15 on October 1, 1989, with most of the F-15A/Bs coming from the disbanded 318th FIS at McChord AFB. Interestingly, the unit's assigned mission is to provide air defense of the U.S. west coast, the same role previously assigned to the 318th.

142 FIG (OR ANG)

F-4C (Mostly Received from Illinois ANG)

63-7490	63-7436	63-7468	63-7475	63-7490	63-7495
63-7517	63-7520#	63-7530	63-7557	63-7570	63-7575
63-7595	63-7602	63-7626	63-7630	63-7647★	63-7670
63-7676#	63-7679	63-7683★	63-7688	63-7699★	63-7707
64-0658	64-0679★	64-0694	64-0699★	64-0711	64-0713
64-0758	64-0766	64-0770	64-0776★★★	64-0777★	64-0794
64-0811	64-0823	64-0825	64-0852	64-0893	64-0926

★ 63-7647 - MiG-17 - June 5, 1967
★ 63-7683 - MiG-21 - January 2, 1967
★ 63-7699 - MiG-17 - May 14, 1967
★ 64-0679 - MiG-17 - July 10, 1965
★ 64-0699 - MiG-17 - April 23, 1966
★★★ 64-0776 - MiG-21 - April 23, 1967 - two - MiG-21 - May 22, 1967
★ 64-0777 - MiG-21 - May 20, 1967
63-7520 Destroyed in a mishap on January 26, 1983
63-7676 Destroyed in a mishap on January 3, 1989

F-4C 63-7672 of the 142nd FIG Red Hawks is seen here in flight . (Boeing)

MiG Killer F-4C 63-7699 is seen here on June 12, 1981 in standard Southeast Asia Camouflage scheme. The only unit markings applied at this time is OREGON in a dark blue stripe at the top of the vertical stabilizer. (Marty Isham)

F-4C 63-7699, on May 14, 1967 while assigned to the 366th TFW, was credited with shooting down a North Vietnamese MiG-17. 7699 was put on display at the Midland Air Museum, England in July 1993. (Ray Leader)

MiG Killer F-4C 64-0776 is seen here on June 12, 1981 in standard Southeast Asia Camouflage scheme with OREGON in a yellow stripe at the top of the vertical stabilizer. 7776 was put on display at the Museum of Flight , Seattle, Washington in October 1989. (Marty Isham)

F-4C 64-0776, seen here on June 29, 1989, was responsible for the downing of three North Vietnamese MiG-17 fighters. The first MiG-17 was shot down on May 12, 1966 while 0660 was assigned to the 35th TFW. The second and third MiG-17 were shot down while 0660 was assigned to the 366th TFW, on May 14, 1967 and June 5, 1967. (Ray Leader)

MiG Killer F-4C 64-0776 is seen here in October 1986. (Jerry Geer)

F-4C 63-7683, seen here in August 1987, while assigned to the 8th TFW, was credited with shooting down a North Vietnamese MiG-21 on January 2, 1967. (Alec Fushi)

MiG Killer F-4C 63-7683 seen here on October 15, 1988 was put on display at the Fuerza Arerea de Chile Museum, Chile in March 1995. (David F. Brown Collection)

F-4C 64-0777, on May 20, 1967 while assigned to the 366th TFW, was credited with shooting down a North Vietnamese MiG-21. 0777 was put on display at the Frontiers of Flight Museum in Dallas, Texas in December 1994. (Alec Fushi)

F-4C 63-7630 is seen here in Air Defense gray and full color Red Hawks markings.The aircraft has a non-standard gray radome. 7630 was retired to Shaw AFB, South Carolina and used as a battle damage repair aircraft and then scrapped (Alec Fushi)

F-4C 64-0658 is seen here on April 16, 1983 in Air Defense gray and full color Red Hawks markings. (Donald S. McGarry)

F-4C 63-7517 is seen here on December 21, 1986 in Hill Grays and subdued gray Red Hawks markings. (Douglas E. Slowiak/Vortex Photo Graphics)

F-4C 63-7575 is seen here in August 1989 in Europe I and subdued black Red Hawks markings. 7575 was later used as a target on the Holloman AFB, New Mexico range complex. (Alec Fushi)

111th FIS, 147th FIG, Ellington Air National Guard Station, Houston, Texas

The 111th FIS "Texans" operated F-4C and F-4D aircraft in an Air Defense role from Ellington ANGS, Houston, Texas 1982 through 1989. The 111th FIS began as the 111th Observation Squadron, established in the Texas National Guard in June 1923. During World War II the 111th flew Douglas A-20 Havocs in Algeria and North American A-36s, F-51s and F-6s in Italy and France. The 111th was reorganized as a National Guard fighter squadron in 1947. Called up for action in Korea with F-84Es, the Texans flew combat missions with between September 1951 and July 1952. After returning home to Houston in 1952, the 111th flew F-51Hs, F-80Cs, F-86D/Ls, a mix of F/TF-102As and F-101B/Fs plus T-33As, and then F-101B/Fs.

In 1982 they received F-4Cs mostly from the 58th TFW at Luke AFB. In 1987 they traded their F-4Cs for F-4Ds from the 31st TFW at Homestead AFB and the 184th TFTG, Kansas ANG at McConnell AFB. The 111th's F-4Ds stood Air Defense alert at Holloman AFB, New Mexico until the squadron began its conversion to F-16.A/Bs in 1989. Its Fighting Falcons were upgraded to the ADF configuration. In late 1996 the ADF Falcons were replaced with Block 25 F-16C/Ds.

147 FIG (TX ANG)

F-4C (Received from Luke AFB)

63-7420	63-7468	63-7484	63-7500	63-7501	63-7517
63-7584	63-7585	63-7589*	63-7591	63-7624	63-7631
63-7650*	63-7685	63-7702	64-0699*	64-0705	64-0712
64-0802	64-0816	64-0828	64-0865	64-0908	64-0911
64-0926	64-0928				

* 63-7589 - MiG-21 - January 2, 1967
* 63-7650 - MiG-17 - September 16, 1966
* 64-0699 - MiG-17 - April 23, 1966

F-4D (Received from Homestead AFB and Kansas ANG)

65-0626	65-0635	65-0641	65-0644	65-0655	65-0660
65-0665	65-0666	65-0672	65-0676	65-0700	65-0707
65-0718	65-0721	65-0735	65-0736	65-0749	65-0758
65-0759	65-0777	65-0801*	66-0261	66-0280*	66-7460
66-7467	66-7500	66-7640	66-7664	66-7708	

* 65-0801 - MiG-21 - May 31, 1972
* 66-0280 - MiG-21 - April 16, 1972

Above: 66-7460 is seen here on December 4, 1989 painted as the 147th FIG Commander's aircraft. The "new" 111th FIS Commander's aircraft, an F-16 (81-0797) can be seen in the background. (Keith Snyder)

Right: As evidenced by this photo, F-4D 66-0261 was marked as the 111th FIS Commander's aircraft. (Pete Wilson)

Below: Newly arrived F-4C 64-0908 is seen here in November 1983. The only unit markings on the aircraft are on the 370 gallon under-wing fuel tanks. (Scott Wilson)

F-4D 65-0655 is seen here in October 1987 in Europe 1 with standard TAC style serial number presentation, but no tail markings. (Don Logan Collection)

F-4D 65-0700 is seen here in October 1987 in Europe I Camouflage and full 147th FIG markings. (David F. Brown Collection)

F-4C 63-7500, marked as 67500 is seen here in Europe I Camouflage on December 27, 1987. 7500 was later used as a target aircraft on the Nellis AFB, Nevada range complex. (Keith Snyder)

F-4D 65-0644 is seen here at Luke AFB, Arizona in Hill Grays paint scheme on May 20, 1989. (Douglas E. Slowiak/Vortex Photo Graphics)

F-4D 66-7467, marked as 67467 is seen here in Hill Grays in November 1987. (David F. Brown Collection)

F-4D 65-0721 is seen here in Hill Grays paint scheme in November 1987. (David F. Brown Collection)

F-4C 64-0705, marked as 40705 is seen here in Air Defense Gray in March 1985. (David F. Brown)

F-4C 64-0911, marked as 40911 is seen here in Air Defense Gray in March 1985. 0911 was put on display at the Pekin Municipal Airport, Pekin, Illinois in November 1991. (David F. Brown)

F-4C 64-0865 is seen here in Air Defense Gray in May 1983. (David F. Brown)

182nd TFS, 149th TFG, Kelly AFB, San Antonio, Texas

Based in San Antonio, Texas, at Kelly AFB, the 182nd flew F-4Cs from spring 1979 through summer of 1986. It had originally been activated as the 182nd Fighter Squadron (SE) in October 1947. It was called to active duty in October 1950 as 182nd Fighter-Bomber Squadron and converted from F-51Ds to F-84Es. It deployed to Korea and flew combat missions between August 1951 and July 1952. After returning home the unit flew F-51 Hs and F-80Cs as a fighter-bomber squadron and moved from Brooks AFB to Kelly AFB in August 1956. It was redesignated the 182nd FIS in January 1957, flying F-80Cs, F-86D/Ls (from December 1957) and F-102A/TF-102As (from summer 1960) in an Air Defense role. It Converting to F-84Fs, it was redesignated 182nd TFS in September 1949. The 182nd continued as a TFS, re-equipped with F-100D/Fs in spring 1971.

In spring 1979 the 182nd received F-4Cs from 401st TFW, Torrejon AB, Spain; Spangdahlem AB, Germany; George AFB; and Luke AFB. The 182nd TFS first deployed outside the North American continent in August 1981 when it sent F-4Cs and crews to Iceland for Coronet Cruise. In the summer of 1986, the 182nd TFS converted to F-16A/Bs. It was redesignated the 182nd Fighter Squadron on March 16, 1992 and currently flies F-16C/Ds.

149 TFG (TX ANG)

F-4C (Received from Torrejon AB, Spangdahlem AB, George AFB, and Luke AFB)

63-7419	63-7421	63-7426	63-7431	63-7439	63-7449
63-7485	63-7487	63-7492	63-7515	63-7550	63-7557
63-7569	63-7585	63-7601	63-7625	63-7638	63-7689
63-7711	64-0750	64-0818	64-0827	64-0829**	64-0879
64-0904	64-0917	64-0918			

** 64-0829 - two - MiG-17 - May 20, 1967

Double MiG Killer F-4C 64-0829 is seen here in August 1983 in Southeast Asia Wraparound Camouflage. (David F. Brown)

F-4C 64-0829 is seen here in Europe I Camouflage in July 1986. While assigned to the 8th TFW at Ubon RTAFB in Thailand, 0829 was responsible for the downing of two North Vietnamese MiG-17s. Both shoot-downs occurred on May 20, 1967 with Col. Robin Olds at the controls. (Tom Brewer Collection)

F-4C 63-7689 is seen here on April 24, 1984 wearing special temporary gray paint for an exercise at Nellis AFB. (David F. Brown)

F-4C 64-0918 is seen here on April 24, 1984 wearing different special temporary gray paint for the same Nellis AFB exercise. (David F. Brown)

F-4C 63-7419 is seen here on June 13, 1979 in standard Southeast Asia Camouflage with the red and white TEXAS tail stripe and white serial number presentation. 7419 was put on display at the RAF Lakenheath Display in February 1995. (Robert J. Mills)

F-4C 63-7449 is seen here on May 3, 1983 in standard Southeast Asia Camouflage with the red and white TEXAS tail stripe and black serial number presentation. 7449 was retired to RAF Upper Heyford, used as a battle damage repair aircraft, and then scrapped. (Scott Wilson)

F-4C 63-7585 is seen here on May 3, 1983 in Southeast Asia Wraparound Camouflage with the red and white TEXAS tail stripe and black serial number presentation. 7585 was put on display at Hermeskeil, Germany in April 1991. (Scott Wilson)

F-4C 63-7421 is seen here on January 25, 1986 in Southeast Asia Wraparound Camouflage with standard TEXAS tail stripe. A black SA (San Antonio) tail code has been added. (Donald S. McGarry)

F-4C 63-7431 is seen here on January 25, 1986 in Southeast Asia Wraparound Camouflage with standard TEXAS tail stripe and SA tail code. 7431 remained at Kelly AFB, Texas and was put on display. (Donald S. McGarry)

F-4C 64-0879 is seen here on January 25, 1986 in Southeast Asia Wraparound Camouflage with standard TEXAS tail stripe and SA tail code. 0879 was retired to Hahn Air Base, Germany, used as a battle damage repair aircraft, and then scrapped. (Donald S. McGarry)

F-4C 63-7550 is seen here on January 25, 1986 in Europe I Camouflage with standard TEXAS tail stripe and SA tail code. 7550 was retired to Robins Air Force Base, Georgia, used as a battle damage repair aircraft, and then scrapped. (Donald S. McGarry)

F-4C 64-0750 is seen here at Kelly AFB, Texas on January 25, 1986. (Donald S. McGarry)

F-4C 63-7569 is seen here waiting to takeoff on a training mission. (Donald S. McGarry)

F-4C 64-0750 is seen here on May 19, 1985. (William R. Peake)

F-4C 64-0818 is seen here on June 2, 1982 with no tail stripe. 0818 was retired to Hill Air Force Base, Utah, used as a battle damage repair aircraft, and then scrapped. (Pete Wilson)

134th TFS, 158th TFG, Burlington Air National Guard Station, Vermont

Located at Burlington ANGS, Vermont, the 134th TFS Green Mountain Boys began conversion to F-4Ds in January 1982. This unit of the Vermont ANG was activated in July 1946 as the 134th Fighter Squadron (SE) and was initially equipped with P-47Ds. It was redesignated the 134th Fighter Interceptor Squadron in 1951, successively flying F-51Ds, F-94A/Bs, F-89Ds and F-102A/TF-102As. In June 1974, it was redesignated 134th Defense Systems Evaluation Squadron in June 1974 it converted from F-102 Delta Daggers to Martin EB-57s.

With its conversion to F-4D received from the 56th TFTW at MacDill AFB in January 1982, the squadron was redesignated the 134th TFS. Converting to F-16 Fighting Falcons beginning in Apri1 1986, the unit was once again designated 134th FIS on July 1, 1987. Redesignated as the 134th FS in March 1992, it began flying F-16A/B ADFs.

158 TFG (VT ANG)

F-4D (Received from MacDill AFB)

65-0608*	65-0617	65-0688	65-0704	65-0730	65-0734
65-0737	65-0748	65-0774	65-0788	65-0790	65-0793
65-0794	66-0226	66-0243	66-0254	66-0266	66-0283
66-7457	66-7464	66-7470	66-7471		

* 65-0608 - MiG-21 - September 12, 1972

Right: The 134th FIS, 158th FIG, "THE GREEN MOUNTAIN BOYS" carried VT (VermonT) tail codes. In these days of political correctness, the unit was able to keep "BOYS" in their nickname (the 174th TFW "BOYS FROM SYRACUSE" were forced to drop their nickname) only because of the historic significance of the "GREEN MOUNTAIN BOYS" with a history all the way back to the American Revolution and the Vermont Militia under Ethan Allen. (David F. Brown Collection)

Below: F-4D 66-7457 is seen here without the yellow Vermont tail stripe. The serial number as applied to this aircraft could be confusing as 66-0457 is an RF-4C. 7457 was retired to Holloman AFB, New Mexico in September 1998, and used as a range target. (Don Logan Collection)

F-4D 65-0704 is seen here at Burlington, Vermont on August 7, 1982 in full Vermont ANG markings. (Marty Isham)

F-4D 65-0734 is seen here on May 22, 1985. The self-contained crew boarding ladder is extended, visible aft of the nose landing gear. (David F. Brown Collection)

F-4D 65-0737 is seen here in July 1985 in full Vermont ANG markings. An eye has been painted on the forward fuselage just aft of the radome. (David F. Brown Collection)

F-4D 65-0793 is seen here on the ramp at Burlington, Vermont in July 1985. (David F. Brown Collection)

F-4D 65-0704 is seen here in May 1985 in Europe I Camouflage, also having the eye on the forward fuselage. (David F. Brown Collection)

F-4D 66-0226 is seen here in July 1985 in Europe I Camouflage. (David F. Brown Collection)

In this view of F-4D 66-0226 on June 10, 1984 the eye has been painted on the forward fuselage. (David F. Brown Collection)

Paint Schemes

The first Air Force Phantoms were delivered in the Navy scheme of Gull Gray upper colors and White undersides. As the F-4s began to see combat in Vietnam, the paint scheme was changed to the Southeast Asia (SEA) scheme of two greens and a tan upper with light gray lower colors. This was the scheme carried by most F-4s when they entered AFRES and ANG service. As experience in 1976 Red Flag missions showed; when the aircraft rolled in for a ground attack, the light gray underside made the aircraft quite visible from above. Based on this, the light gray was deleted and the upper colors wrapped around to the bottom.

Phantoms used for Air Defense roles were painted a gloss medium gray known as ADC gray also used on other Interceptors such as F-101s, F-102s, and F-106s.

In the late 1970s attention returned to Western Europe where the lighter tan color was completely inappropriate for the terrain. It was replaced with a dark gray creating a scheme known as European 1.

The final paint schemes used on Phantoms were a series of gray schemes known as Hill Gray I and Hill Gray II. They were named after Hill AFB, the Air Logistics Center for the F-4. These two schemes were determined to be the best to prevent easy visual detection. The Hill Gray I consisted of three grays and was in limited use at the beginning of the program. The two gray Hill Gray II or Wraparound Gray scheme as it was also called was the final scheme used by Phantoms in the AFRES and ANG.

Two units – the 704th TFS AFRES at Carswell AFB and the 122nd TFS, Louisiana ANG at NAS New Orleans used distinctive schemes developed by the units themselves.

Following are a listing of standard F-4 paint schemes in order of use. Also listed are the Federal Standard (FS) colors for that particular scheme.

Southeast Asia (SEA) Scheme
Forest Green (FS 34079)
Medium Green (FS 34102)
Tan (FS 30219)
Camouflage Gray (FS 36622)

SEA Wraparound Scheme
Forest Green (FS 34079)
Medium Green (FS 34102)
Tan (FS 30219)

Air Defense Command (ADC) Gray Scheme
ADC Gray (FS 16473)

European 1 Scheme
Forest Green (FS 34079)
Medium Green (FS 34102)
Euro 1 Gray (FS 36081)

Hill Gray I Scheme
Gunship Gray (FS 36118)
Light Compass Gray (FS 36375)
Neutral Gray (FS 36270)

Hill Gray II Scheme
Gunship Gray (FS 36118)
Neutral Gray (FS 36270)

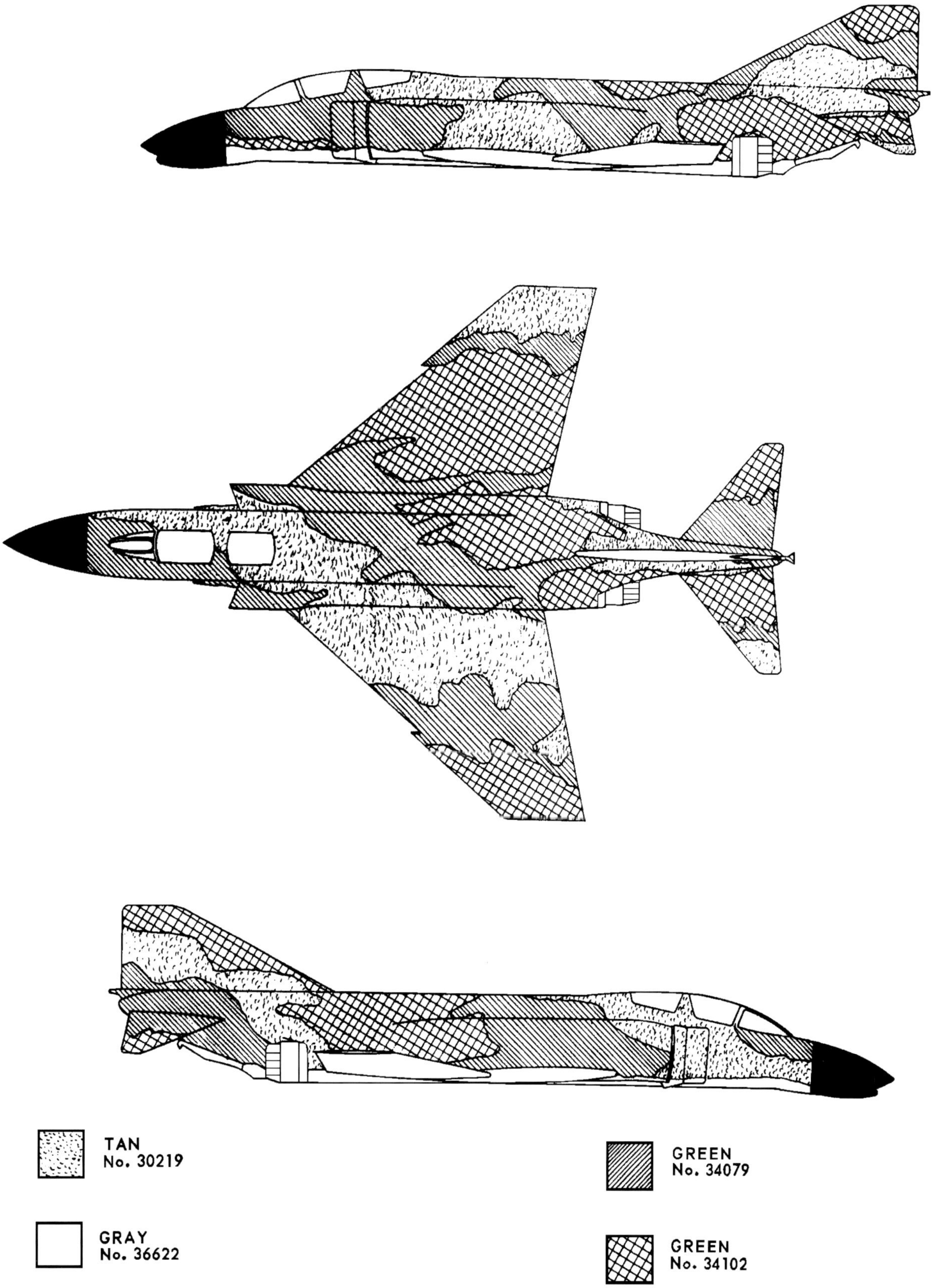
TAN
No. 30219
GREEN
No. 34079
GRAY
No. 36622
GREEN
No. 34102

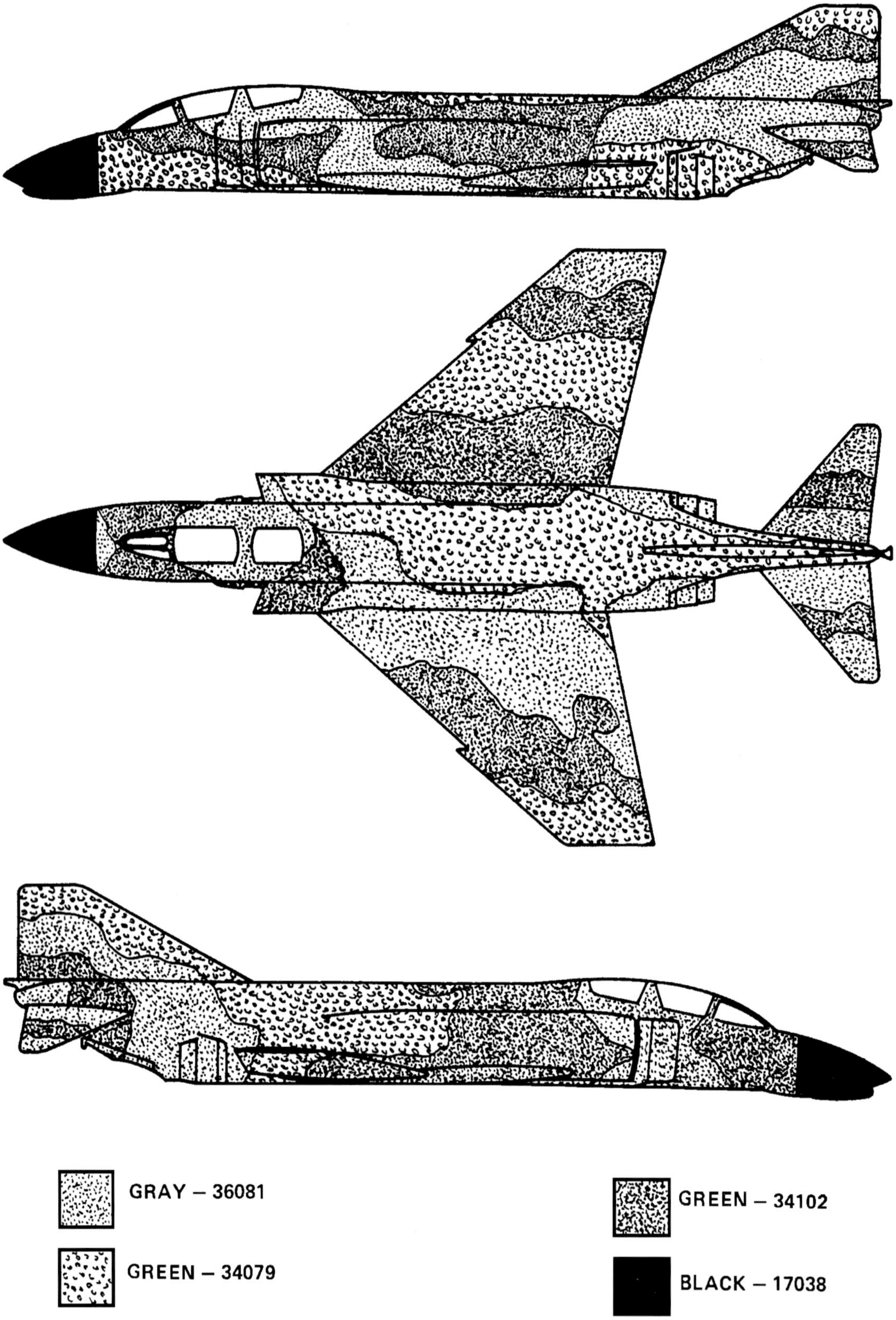
GRAY – 36081
GREEN – 34079
GREEN – 34102
BLACK – 17038

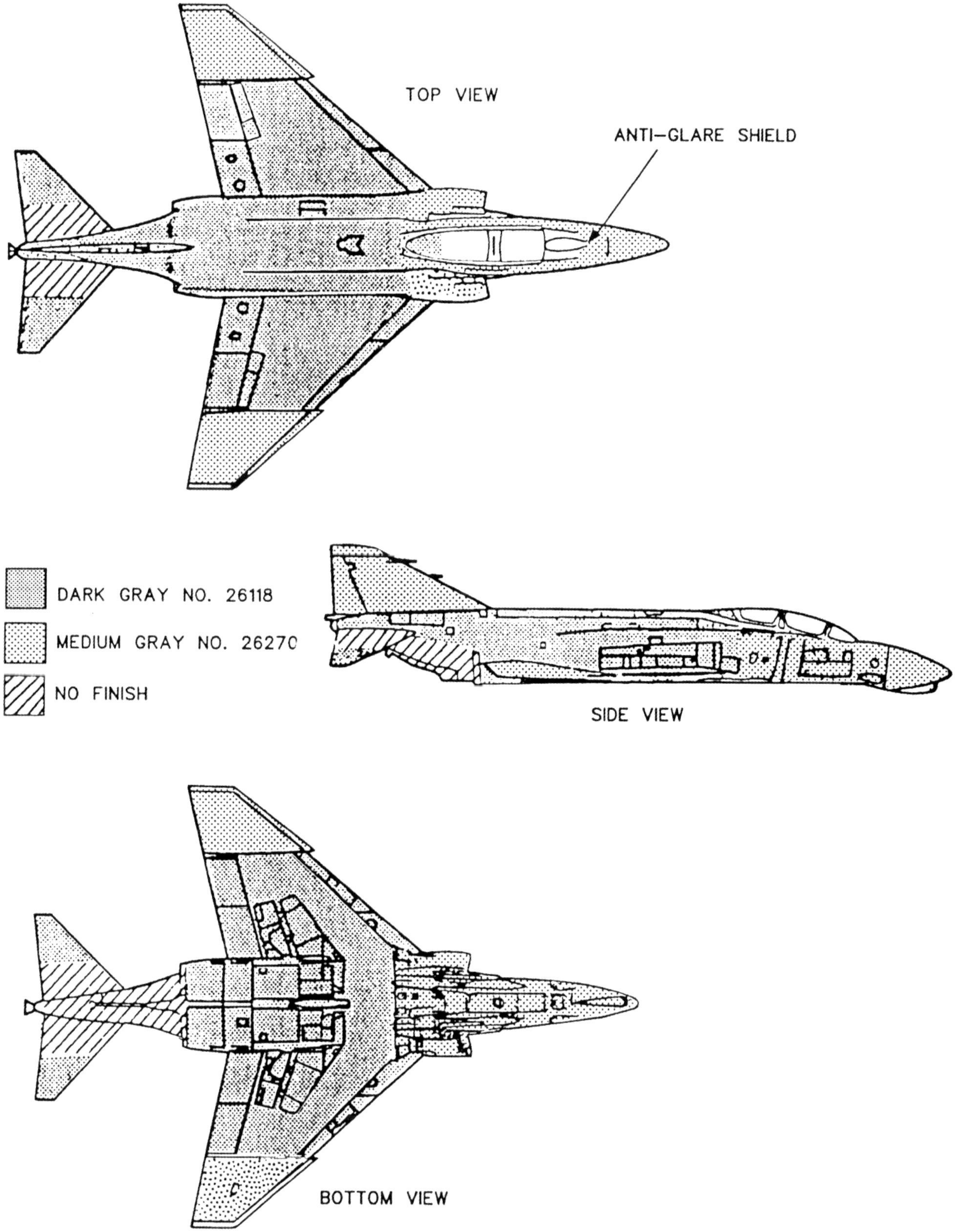
TOP VIEW
ANTI-GLARE SHIELD
DARK GRAY NO. 26118
MEDIUM GRAY NO. 26270
NO FINISH
SIDE VIEW
BOTTOM VIEW

Navigation and Weapons Delivery System (NWDS) AN/ASQ-203

Early in 1987, the Boeing Military Airplane Company in Wichita Kansas started work on the AN/ASQ-203 Navigation and Weapon Delivery System (NWDS) for the F-4. The purpose was to replace the existing F-4 navigation and weapon delivery system in non AN/ARN-101 DMAS F-4s. A version for the RF-4C, F-4E, and F-4G was being designed. This NWDS would replace the existing mechanical gyroscope based Inertial Navigation System with a new Ring-LASER gyro. The system was highly reliable and designed to reduce maintenance cost per flight hour. By April 1988 system design on the system for the RF-4C and the F-4E had progressed to the hardware phase. By early 1989 the AN/ASQ-203(V)1 for the RF-4C and the AN/ASQ-203(V)2 for the F-4E had entered production. NWDS was installed on selected RF-4C aircraft by TCTO 1F-4(R)C-706 beginning in September 1989. The system was operational in RF-4Cs of the Nevada Air National guard in Reno and RF-4Cs of the Alabama Air National Guard in Birmingham by the end of 1990. NWDS jets from these units took part in Desert Storm. The F-4E NWDS was fielded by the Indiana ANG using TCTO 1F-4-671 beginning in late 1989.The version for the F-4G was postponed and finally canceled due to the belief the F-4Gs would be retired from the inventory before the cost of the F-4G NWDS could be recovered.

The F-4Es equipped with NWDS were used only by the Indiana Air National Guard, by both the 113th TFS, 181st TFG at Terre Haute and 163rd TFS, 122nd TFW at Fort Wayne. The NWDS F-4Es not retired to AMARC were sold to GREECE under FMS coming from the 122nd TFW at Fort Wayne. Additional F-4E kits were sold to the German Air Force and were used in their Training program at George AFB, California and later at Holloman AFB, New Mexico.

Reconnaissance Squadrons in the Alabama, California, Idaho, Mississippi, Nebraska, and Nevada Air National Guards operated NWDS RF-4Cs. The RF-4C NWDS was similar to the F-4E system except it added a Pilots Auxiliary Display (PAD) which was a two line alphanumeric display capable of displaying selected lines of navigation and reconnaissance system data to the pilot.

NWDS

The NWDS is an integrated modular navigation, weapon delivery, reconnaissance (RF-4 only), and pathfinder (RF-4 only) system installed in the RF-4C [AN/ASQ-203(V)1] and the F-4E [AN/ASQ-203(V)2]. The system was designed to enhance capabilities of the existing aircraft systems. The NWDS in the RF-4C replaced the existing Navigation Computer (AN/ASN-46), Inertial Navigation System (AN/ASN-56), and Gyro Stability Platform (AN/ASN-74). The NWDS In the F-4E replaced the existing Navigation Computer (AN/ASN-46A), Inertial Navigation System (AN/ASN-63), and Weapon Release Computer Set (AN/ASQ-91).

The functions of the replaced equipment were retained in NWDS with the addition of a Data Transfer Set, which allowed loading of navigation, and weapon data into the NWDS. NWDS had the capability of computing location both in standard Latitude and Longitude coordinate system, and Universal Transverse Mercator (UTM) coordinate system, and converting the coordinates between the two systems. The NWDS system had the capability to add Global Positioning System (GPS) equipment as a growth item, however, GPS was not added prior to retiring of the NWDS aircraft.

F-4E OPERATION

F-4E NWDS had two major operational modes: Navigation and Weapon Delivery. NWDS used displays on a small computer screen on the CDU (Control/Display Unit) for WSO interface with the navigation and weapon delivery functions. In the Navigation mode the NWDS provided precision navigation to and from the target area using 99 stored Destinations with their associated Cue Points, up to 10 Freeze Points, and up to 16 Weapon Points for display, steering and recall. During Weapon Delivery, precision navigation and steering functions allowed delivery of a variety of weapon types using both computer aided and manual modes. The NWDS also provided an interface for both Electro-Optical and Infrared AGM-65 Maverick missiles. The AN/AVQ-23 Pave Spike Laser Designator Pod was integrated with the NWDS. During Weapon Delivery modes, the NWDS computed weapon release points based on data loaded in the system automatically by the Data Transfer Set, or manually by the WSO, for the type weapon to be delivered. In both Navigation and Weapon Delivery modes, the NWDS could cue sensors to the destination or offset points for navigation updates or secondary targets. The navigation system could be updated in both altitude and position. Lead Computing Optical Sight System (LCOSS), Pave Spike, Radar, TACAN, and overfly provided position data. Pave Spike, A/G Radar, Combined Altitude Radar Altimeter (CARA), Manual, and Barometric Altitude provided altitude update data. In addition NWDS provided a self-contained approach to landing mode.

RF-4C OPERATION

RF-4C NWDS had two major operational modes: Navigation and Reconnaissance. In the Navigation mode the NWDS provided precision navigation to and from the reconnaissance or target area. During Reconnaissance, precision navigation and steering functions were provided to allow performance of point target missions, route mapping missions, and mosaic mapping missions which included race track and adjacent parallel reconnaissance profiles. The Navigation system could be updated in both altitude and position. Forward looking radar (AN/APQ-172), TACAN, and Overfly provided position update. Altitude update data was provided by Combined Altitude Radar Altimeter (CARA), Barometric Altitude, and manual altitude input.

F-4 Navigation And Weapons Delivery System (NWDS) AN/ASQ-203

NWDS F-4E JETS - AN/ASQ-203(V)2

67-0345	FMS (GREECE)
67-0348	RETIRED(AMARC)
67-0350	FMS (GREECE)
67-0377	FMS (GREECE)
67-0381	FMS (GREECE)
67-0384	RETIRED
67-0389	DISPLAY-FT.WAYNE, IN
68-0305	RETIRED(AMARC)
68-0318	FMS (GREECE)
68-0321	RETIRED(AMARC)
68-0337	RETIRED(AMARC)
68-0361	FMS (GREECE)
68-0363	FMS (GREECE)
68-0372	RETIRED(AMARC)
68-0381	FMS (GREECE)
68-0393	FMS (GREECE)
68-0394	FMS (GREECE)
68-0402	FMS (GREECE)
68-0405	FMS (GREECE)
68-0408	FMS (GREECE)
68-0411	ATTRITED
68-0412	FMS (GREECE)
68-0418	RETIRED(AMARC)
68-0424	FMS (GREECE)
68-0426	FMS (GREECE)
68-0429	ATTRITED
68-0432	FMS (GREECE)
68-0438	FMS (GREECE)
68-0440	FMS (GREECE)
68-0442	FMS (GREECE)
68-0443	RETIRED(AMARC)
68-0444	FMS (GREECE)
68-0445	FMS (GREECE)
68-0467	RETIRED
68-0478	RETIRED(AMARC)
68-0480	FMS (GREECE)
68-0481	FMS (GREECE)

68-0490	RETIRED(AMARC)
68-0495	RETIRED(AMARC)
68-0496	FMS (GREECE)
68-0503	RETIRED(AMARC)
68-0506	FMS (GREECE)
68-0515	FMS (GREECE)
68-0516	RETIRED(AMARC)
68-0517	FMS (GREECE)
68-0529	RETIRED(AMARC)
68-0531	RETIRED(AMARC)
68-0538	RETIRED(AMARC)
75-0628	RETIRED(AMARC)
75-0629	RETIRED(AMARC)
75-0631	RETIRED(AMARC)
75-0632	RETIRED(AMARC)
75-0633	RETIRED(AMARC)
75-0635	RETIRED(AMARC)
75-0636	RETIRED(AMARC)

RF-4C NWDS JETS AN/ASQ-203(V)1

63-7742	RETIRED(AMARC)
63-7754	RETIRED(AMARC)
64-0997	RETIRED(AMARC)
64-1005	RETIRED(AMARC)
64-1006	FMS (SPAIN)
64-1010	RETIRED(AMARC)
64-1014	RETIRED(AMARC)
64-1016	RETIRED(AMARC)
64-1017	RETIRED(AMARC)
64-1021	RETIRED(AMARC)
64-1022	DISPLAY-SUSANVILLE, CA
64-1023	RETIRED(AMARC)
64-1028	RETIRED(AMARC)
64-1029	RETIRED(AMARC)
64-1030	RETIRED(AMARC)
64-1031	RETIRED(AMARC)
64-1033	RETIRED(AMARC)
64-1034	ATTRITED
64-1035	RETIRED(AMARC)
64-1039	FMS (SPAIN)
64-1041	RETIRED(AMARC)
64-1047	DISPLAY-AF MUSEUM
64-1050	RETIRED(AMARC)
64-1051	RETIRED(AMARC)
64-1055	RETIRED(AMARC)
64-1056	ATTRITED
64-1057	RETIRED(AMARC)
64-1058	RETIRED(AMARC)
64-1060	RETIRED(AMARC)
64-1063	RETIRED(AMARC)
64-1065	RETIRED(AMARC)
64-1068	RETIRED(AMARC)
64-1075	RETIRED(AMARC)
64-1077	RETIRED(AMARC)
64-1079	RETIRED(AMARC)
64-1080	RETIRED(AMARC)
65-0823	FMS (SPAIN)
65-0824	RETIRED(AMARC)
65-0828	RETIRED(AMARC)
65-0832	RETIRED(AMARC)
65-0833	DISPLAY-JASPER, AL
65-0838	RETIRED(AMARC)
65-0843	RETIRED(AMARC)
65-0852	RETIRED(AMARC)
65-0853	RETIRED(AMARC)
65-0854	RETIRED(AMARC)
65-0859	RETIRED(AMARC)
65-0864	FMS (SPAIN)
65-0867	RETIRED(AMARC)
65-0870	RETIRED(AMARC)
65-0874	RETIRED(AMARC)
65-0875	RETIRED(AMARC)
65-0876	FMS (SPAIN)
65-0879	RETIRED(AMARC)
65-0884	RETIRED(AMARC)
65-0886	DISPLAY-RENO,NV
65-0890	RETIRED(AMARC)
65-0892	RETIRED(AMARC)
65-0897	FMS (SPAIN)
65-0898	RETIRED(AMARC)
65-0900	CRASHED
65-0901	RETIRED(AMARC)
65-0902	RETIRED(AMARC)
65-0903	DISPLAY- COLUMBUS,OH
65-0907	RETIRED(AMARC)
65-0908	RETIRED(AMARC)
65-0910	RETIRED(AMARC)
65-0911	RETIRED(AMARC)
65-0912	RETIRED(AMARC)
65-0917	RETIRED(AMARC)
65-0918	RETIRED(AMARC)
65-0922	RETIRED(AMARC)
65-0923	RETIRED(AMARC)
65-0928	RETIRED(AMARC)
65-0931	RETIRED(AMARC)
65-0932	RETIRED(AMARC)
65-0935	RETIRED(AMARC)
65-0939	RETIRED(AMARC)
65-0942	RETIRED(AMARC)
65-0945	RETIRED(AMARC)

66-0397	ABDR(NELLIS)
66-0399	RETIRED(AMARC)
66-0400	RETIRED(AMARC)
66-0401	RETIRED(AMARC)
66-0402	RETIRED(AMARC)
66-0407	RETIRED(AMARC)
66-0408	RETIRED(AMARC)
66-0415	RETIRED(AMARC)
66-0417	RETIRED(AMARC)
66-0418	RETIRED(AMARC)
66-0422	RETIRED(AMARC)
66-0423	RETIRED(AMARC)
66-0425	RETIRED(AMARC)
66-0428	RETIRED(AMARC)
66-0430	RETIRED(AMARC)
66-0435	RETIRED(AMARC)
66-0444	RETIRED(AMARC)
66-0449	RETIRED(AMARC)
66-0453	RETIRED(AMARC)
66-0464	RETIRED
66-0469	DISPLAY-HILL AFB, UT
67-0434	RETIRED(AMARC)
67-0449	RETIRED(AMARC)
67-0452	DISPLAY-EGLIN AFB, FL
67-0456	RETIRED(AMARC)
67-0463	RETIRED(AMARC)
67-0466	RETIRED(AMARC)
68-0575	RETIRED(AMARC)
68-0593	RETIRED(AMARC)

F-4C & F-4D SERIAL NUMBERS

BLOCK NUMBERS

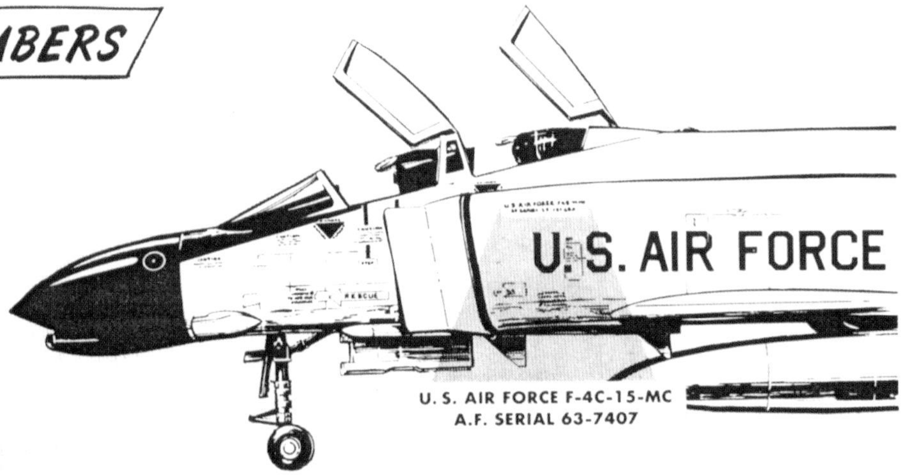

U. S. AIR FORCE F-4C-15-MC
A.F. SERIAL 63-7407

F-4C-15-MC
A.F. SERIAL 62-12199 and
A.F. SERIAL 63-7407 thru 63-7420

F-4C-16-MC
A.F. SERIAL 63-7421 thru 63-7442

F-4C-17-MC
A.F. SERIAL 63-7443 thru 63-7468

F-4C-18-MC
A.F. SERIAL 63-7469 thru 63-7526

F-4C-19-MC
A.F. SERIAL 63-7527 thru 63-7597

F-4C-20-MC
A.F. SERIAL 63-7598 thru 63-7662

F-4C-21-MC
A.F. SERIAL 63-7663 thru 63-7713
and 64-654 thru 64-672

F-4C-22-MC
A.F. SERIAL 64-673 thru 64-737

F-4C-23-MC
A.F. SERIAL 64-738 thru 64-817

F-4C-24-MC
A.F. SERIAL 64-818 thru 64-881

F-4C-25-MC
A.F. SERIAL 64-882 thru 64-928

F-4D-24-MC
A.F. SERIAL 64-929 thru 64-937

F-4D-25-MC
A.F. SERIAL 64-938 thru 64-963

F-4D-26-MC
A.F. SERIAL 64-964 thru 64-980
and 65-580 thru 65-611

F-4D-27-MC
A.F. SERIAL 65-612 thru 65-665

F-4D-28-MC
A.F. SERIAL 65-666 thru 65-770

F-4D-29-MC
A.F. SERIAL 65-771 thru 65-801 and
66-226 thru 66-283 and
66-7455 thru 66-7504

F-4D-30-MC
A.F. SERIAL 66-7505 thru 66-7650

F-4D-31-MC
A.F. SERIAL 66-7651 thru 66-7774
and 66-8685 thru 66-8698

F-4D-32-MC
A.F. SERIAL 66-8699 thru 66-8786

F-4D-33-MC
A.F. SERIAL 66-8787 thru 66-8825

RF-4C SERIAL NUMBERS

BLOCK NUMBERS

U.S. AIR FORCE RF-4C-17-MC
A.F. SERIAL 63-7742

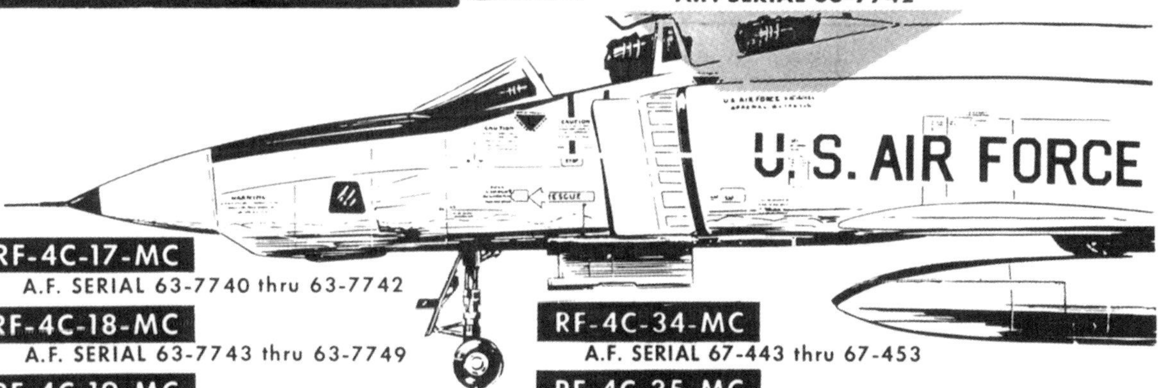

RF-4C-17-MC
A.F. SERIAL 63-7740 thru 63-7742

RF-4C-18-MC
A.F. SERIAL 63-7743 thru 63-7749

RF-4C-19-MC
A.F. SERIAL 63-7750 thru 63-7763

RF-4C-20-MC
A.F. SERIAL 64-997 thru 64-1017

RF-4C-21-MC
A.F. SERIAL 64-1018 thru 64-1037

RF-4C-22-MC
A.F. SERIAL 64-1038 thru 64-1061

RF-4C-23-MC
A.F. SERIAL 64-1062 thru 64-1077

RF-4C-24-MC
A.F. SERIAL 64-1078 thru 64-1085
and 65-818 thru 65-838

RF-4C-25-MC
A.F. SERIAL 65-839 thru 65-864

RF-4C-26-MC
A.F. SERIAL 65-865 thru 65-901

RF-4C-27-MC
A.F. SERIAL 65-902 thru 65-932

RF-4C-28-MC
A.F. SERIAL 65-933 thru 65-945
and 66-384 thru 66-388

RF-4C-29-MC
A.F. SERIAL 66-387 thru 66-406

RF-4C-30-MC
A.F. SERIAL 66-407 thru 66-428

RF-4C-31-MC
A.F. SERIAL 66-429 thru 66-450

RF-4C-32-MC
A.F. SERIAL 66-451 thru 66-472

RF-4C-33-MC
A.F. SERIAL 66-473 thru 66-478
and 67-428 thru 67-442

RF-4C-34-MC
A.F. SERIAL 67-443 thru 67-453

RF-4C-35-MC
A.F. SERIAL 67-454 thru 67-461

RF-4C-36-MC
A.F. SERIAL 67-462 thru 67-469

RF-4C-37-MC
A.F. SERIAL 68-548 thru 68-561

RF-4C-38-MC
A.F. SERIAL 68-562 thru 68-576

RF-4C-39-MC
A.F. SERIAL 68-577 thru 68-593

RF-4C-40-MC
A.F. SERIAL 68-594 thru 68-611

RF-4C-41-MC
A.F. SERIAL 69-349 thru 69-357

RF-4C-42-MC
A.F. SERIAL 69-358 thru 69-366

RF-4C-43-MC
A.F. SERIAL 69-367 thru 69-375

RF-4C-44-MC
A.F. SERIAL 69-376 thru 69-384

RF-4C-48-MC
A.F. SERIAL 71-248 thru 71-252

RF-4C-49-MC
A.F. SERIAL 71-253 thru 71-259

RF-4C-51-MC
A.F. SERIAL 72-145 thru 72-150

RF-4C-52-MC
A.F. SERIAL 72-151 thru 72-153

RF-4C-53-MC
A.F. SERIAL 72-154 thru 72-156

F-4E SERIAL NUMBERS

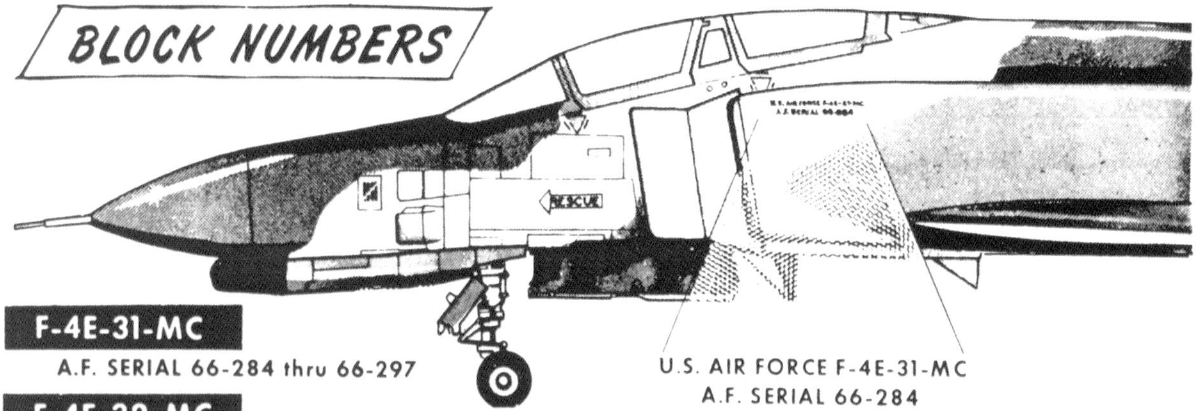

F-4E-31-MC
A.F. SERIAL 66-284 thru 66-297

F-4E-32-MC
A.F. SERIAL 66-298 thru 66-338

F-4E-33-MC
A.F. SERIAL 66-339 thru 66-382
A.F. SERIAL 67-208 thru 67-219

F-4E-34-MC
A.F. SERIAL 67-220 thru 67-282

F-4E-35-MC
A.F. SERIAL 67-283 thru 67-341

F-4E-36-MC
A.F. SERIAL 67-342 thru 67-398

F-4E-37-MC
A.F. SERIAL 68-303 thru 68-365

F-4E-38-MC
A.F. SERIAL 68-366 thru 68-395
A.F. SERIAL 68-400 thru 68-409

F-4E-39-MC
A.F. SERIAL 68-410 thru 68-413
A.F. SERIAL 68-418 thru 68-433
A.F. SERIAL 68-438 thru 68-451

F-4E-40-MC
A.F. SERIAL 68-452 thru 68-453
A.F. SERIAL 68-458 thru 68-468
A.F. SERIAL 68-473 thru 68-483
A.F. SERIAL 68-488 thru 68-494

F-4E-41-MC
A.F. SERIAL 68-495 thru 68-498
A.F. SERIAL 68-503 thru 68-518
A.F. SERIAL 68-526 thru 68-538

F-4E-42-MC
A.F. SERIAL 69-236 thru 69-303

F-4E-43-MC
A.F. SERIAL 69-304 thru 69-307
A.F. SERIAL 69-7201 thru 69-7260

F-4E-44-MC
A.F. SERIAL 69-7261 thru 69-7303
A.F. SERIAL 69-7546 thru 69-7578

F-4E-45-MC
A.F. SERIAL 69-7579 thru 69-7589

F-4E-48-MC
A.F. SERIAL 71-224 thru 71-247

F-4E-49-MC
A.F. SERIAL 71-1070 thru 71-1093

F-4E-50-MC
A.F. SERIAL 71-1391 thru 71-1402
A.F. SERIAL 72-121 thru 72-138

F-4E-51-MC
A.F. SERIAL 72-139 thru 72-144
A.F. SERIAL 72-157 thru 72-159

F-4E-52-MC
A.F. SERIAL 72-160 thru 72-165

F-4E-53-MC
A.F. SERIAL 72-166 thru 72-168
A.F. SERIAL 72-1407

F-4E-54-MC
A.F. SERIAL 72-1476 thru 72-1489

F-4E-55-MC
A.F. SERIAL 72-1490 thru 72-1497

F-4E-56-MC
A.F. SERIAL 72-1498 thru 72-1499

F-4E-57-MC
A.F. SERIAL 73-01157 thru 73-01164

F-4E-58-MC
A.F SERIAL 73-01165 thru 73-01184

F-4E-59-MC
A.F. SERIAL 73-01185 thru 73-01204

F-4E-60-MC
A.F. SERIAL 74-00643 thru 74-00666
A.F. SERIAL 74-01038 thru 74-01049

F-4E-61-MC
A.F. SERIAL 74-01050 thru 74-01061
A.F. SERIAL 74-01620 thru 74-01637

F-4E-62-MC
A.F. SERIAL 74-01638 thru 74-01653

F-4E-63-MC
A.F. SERIAL 75-00628 thru 75-00637

F-4G SERIAL NUMBERS

BLOCK NUMBERS

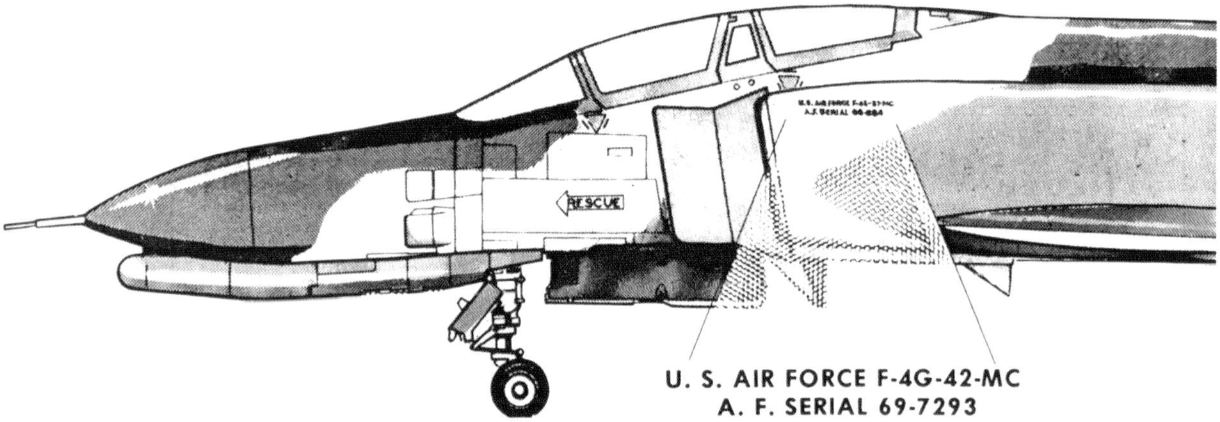

F-4G-42-MC

A.F. SERIAL 69-0236 thru 69-0255
A.F. SERIAL 69-0258 thru 69-0261
A.F. SERIAL 69-0263 thru 69-0265
A.F. SERIAL 69-0267
A.F. SERIAL 69-0269 thru 69-0275
A.F. SERIAL 69-0277 thru 69-0281
A.F. SERIAL 69-0284 thru 69-0286
A.F. SERIAL 69-0290 thru 69-0293
A.F. SERIAL 69-0297

F-4G-43-MC

A.F. SERIAL 69-0303 thru 69-0307
A.F. SERIAL 69-7201 thru 69-7202
A.F. SERIAL 69-7204
A.F. SERIAL 69-7206 thru 69-7212
A.F. SERIAL 69-7214 thru 69-7220
A.F. SERIAL 69-7223
A.F. SERIAL 69-7228
A.F. SERIAL 69-7231 thru 69-7236
A.F. SERIAL 69-7251 thru 69-7254
A.F. SERIAL 69-7256 thru 69-7258
A.F. SERIAL 69-7260

F-4G-44-MC

A.F. SERIAL 69-7261 thru 69-7263
A.F. SERIAL 69-7267
A.F. SERIAL 69-7270
A.F. SERIAL 69-7272
A.F. SERIAL 69-7286 thru 69-7291
A.F. SERIAL 69-7293 thru 69-7295
A.F. SERIAL 69-7297 thru 69-7298
A.F. SERIAL 69-7300 thru 69-7303
A.F. SERIAL 69-7546
A.F. SERIAL 69-7550 thru 69-7551
A.F. SERIAL 69-7556 thru 69-7558
A.F. SERIAL 69-7560 thru 69-7561
A.F. SERIAL 69-7566
A.F. SERIAL 69-7571 thru 69-7572
A.F. SERIAL 69-7574

F-4G-45-MC

A.F. SERIAL 69-7579 thru 69-7583
A.F. SERIAL 69-7586 thru 69-7588

F-4 MODEL MAIN DIFFERENCES

	F-4C	F-4D	F-4E	F-4E AFTER TO 1F-4E-626	F-4G
ENGINES	J79-GE-15	J79-GE-15/15A/15E	J79-GE-17A/E/F/G	J79-GE-17A/E/F/G	J79-GE-17A/C/E/F/G
NO. 7 FUEL CELL	NO	NO	YES	YES	YES
RAM AIR TURBINE	YES	YES	NO	NO	NO
HYDRAULIC WING FOLD	YES	YES	NO	NO	NO
INTERNALLY MOUNTED GUN	NO	NO	YES	YES	NO
RADAR SET	AN/APQ-100	AN/APQ-109	AN/APQ-120	AN/APQ-120 MODIFIED	AN/APQ-120 MODIFIED
INTERCEPT COMPUTER	AN/APA-157	AN/APA-157 or AN/APA-165	AN/APQ-120	AN/APQ-120	AN/APQ-120
OPTICAL SIGHT	FIXED	AN/ASG-22	AN/ASG-26	AN/ASG-26 MODIFIED	AN/APQ-30 MODIFIED
TISEO (AN/ASX-1)	NO	NO	71-237 AND UP	AN/ASX-1 MODIFIED	NO
RADAR RECEIVING SET	NO	NO	NO	NO	AN/APR-47
WEAPONS RELEASE COMPUTER	NONE	AN/ASQ-91	AN/ASQ-91	NO	NO
INERTIAL NAVIGATION SET	AN/ASN-48	AN/ASN-63	AN/ASN-63	NO	NO
NAVIGATION COMPUTER	AN/ASN-46	AN/ASN-46A	AN/ASN-46A	NO	NO
AUXILIARY POWER UNIT	NO	NO	68-452 AND UP	YES	YES
SELF SEALING FUSELAGE FUEL CELLS	NO	NO	68-495 AND UP	YES	YES
LEADING EDGE BOUNDARY LAYER CONTROL	YES	YES	NO	NO	NO
LEADING EDGE SLATS	NO	NO	YES	YES	YES
ATTITUDE REFERENCE BOMBING COMPUTER SET	AN/AJB-7	AN/AJB-7	AN/AJB-7	AN/AJB-7 MODIFIED	AN/AJB-7 MODIFIED
DIGITAL MODULAR AVIONICS SYSTEM (DMAS)	NO	NO	NO	YES	YES
AUTOMATIC FLIGHT CONTROL SYSTEM (AFCS)	AN/ASA-32J	AN/ASA-32J	AN/ASA-32J	AN/ASA-32J MODIFIED	AN/ASA-32J MODIFIED

Air Force Reserve & Air National Guard F-4 Losses

SERIAL NUMBER	MODEL	DATE	UNIT
63-7470	F-4C	November 19,1986	181 TFG, Indiana ANG
63-7481	F-4C	March 3, 1982	122 TFW, Indiana ANG
63-7492	F-4C	January 14, 1986	159 TFG, Louisiana ANG
63-7520	F-4C	January 26, 1983	142 FIG, Oregon ANG, Portland
63-7676	F-4C	January 3. 1989	142 FIG, Oregon ANG, Portland
63-7696	F-4C	June 5, 1985	188 TFG, Arkansas ANG
63-7756	RF-4C	February 2, 1983	187 TRG, Alabama ANG
64-0686	F-4C	February 17, 1984	159 TFG, Louisiana ANG
64-0923	F-4C	March 21, 1987	163 TFG, California ANG
64-1003	RF-4C	November 21, 1980	152 TRG, Nevada ANG
64-1020	RF-4C	April 27, 1988	152 TRG, Nevada ANG
64-1032	RF-4C	June 30, 1985	117 TRW, Alabama ANG
64-1034	RF-4C	February 17, 1993	117 RW, Alabama ANG
64-1036	RF-4C	May 21, 1985	117 TRW, Alabama ANG
64-1044	RF-4C	October 8, 1990	117 TRW, Alabama ANG
64-1056	RF-4C	March 30, 1991	152 TRG, Nevada ANG
64-1073	RF-4C	May 16, 1992	123 TRW, Kentucky ANG
64-1076	RF-4C	May 25, 1978	148 TRG, Minnesota ANG
65-0674	F-4D	July 15, 1983	108 TFW, New Jersey ANG
65-0891	RF-4C	November 14, 1978	148 TRS, Minnesota ANG
65-0900	RF-4C	March 6, 1994	117 RW, Alabama ANG
65-0904	RF-4C	May 17, 1990	124 TRG, Idaho ANG
65-0929	RF-4C	November 18, 1986	124 TRG, Idaho ANG
65-0934	RF-4C	November 20, 1977	155 TRG, Nebraska ANG
66-0413	RF-4C	June 30, 1985	186 TRG, Mississippi ANG
66-7666	F-4D	July 1, 1985	184 TFG, Kansas ANG
66-7675	F-4D	March 5, 1986	183 TFG, Illinois ANG
66-8689	F-4D	September 24, 1983	116 TFW, Georgia ANG
66-8803	F-4D	May 22, 1987	301 TFW, Carswell AFB
66-8823	F-4D	August 21, 1989	507 TFG, Tinker AFB
67-0394	F-4E	July 6, 1989	163 TFG, California ANG

SERIAL NUMBER	MODEL	DATE	UNIT
67-0439	RF-4C	March 26, 1980	186 TRG, Mississippi ANG
68-0311	F-4E	April 8, 1990	131 TFW, Missouri ANG
68-0367	F-4E	July 27, 1989	163 TFG, California ANG
68-0429	F-4E	September 19, 1990	181 TFG, Indiana ANG
68-0448	F-4E	March 12, 1987	131 TFW, Missouri ANG
68-0465	F-4E	February 10, 1989	122 TFW, Indiana ANG
68-0508	F-4E	June 18, 1986	108 TFW, New Jersey ANG
68-0512	F-4E	June 13, 1990	163 TFG, California ANG

NOTES

NOTES

NOTES

NOTES

Also from the Publisher

USAF F-15 EAGLES: Units, Colors & Markings. *Don Logan*. This book chronicles the story of the F-15A, B, C, D, and E. The book includes F-15 program history, prototype and operational aircraft paint schemes, and unit markings of all USAF and Air National Guard units operating the F-15. A section on USAF F-15 MiG killers in both Operation Desert Storm and Operation Allied Force, with photos of all USAF F-15 MiG killers, is included. Over 130 colorful official and unofficial unit patches involving the F-15 are included. The book also includes a description of the aircraft, its systems, and weapons.
Size: 8 1/2" x 11" • over 750 color photographs • 304 pp.
ISBN: 0-7643-1060-7 • hard cover • 59.95

REPUBLIC'S A-10 THUNDERBOLT II: A Pictorial History. *Don Logan*. This book chronicles the story of the A-10, including program history, aircraft paint schemes, all units operating the aircraft and their markings, and selected nose art and ladder door art. Over 100 colorful official and unofficial patches of the units and combat operations involving the A-10 are included. The book also includes a description of the aircraft, its systems, and weapons.
Size: 8 1/2" x 11" • 550+ color photos, line drawings • 240 pp.
ISBN: 0-7643-0147-0 • hard cover • $49.95

THE BOEING C-135 SERIES: Stratotanker, Stratolifter and other Variants. *Don Logan*. This book tells the story of the Boeing C-135 series including: tankers, reconnaissance, airborne command post, weather, test, and special purpose models. All C-135 aircraft types, along with their operating units are covered. Tables and serial number lists are included listing all C-135 configurations by serial number. Also included: a listing of all C-135 losses, including date and reason for loss; three views of C-135 major configurations; selected aircraft nose art; and all USAF, ARFES, and Air National Guard unit markings.
Size: 8 1/2" x 11 • 600+ color photos, line drawings • 256 pp.
ISBN: 0-7643-0286-8 • hard cover • $49.95

NORTHROP'S YF-17 COBRA: A Pictorial History. *Don Logan*. The Northrop YF-17 holds a special place in aircraft history. The YF-17 was one of two prototypes tested in the U.S. Air Force Air Combat Fighter competition, a program which attempted to reverse the trend of increasing cost and complexity of new fighter aircraft, and which resulted in the selection and manufacture of the F-16 as the next generation free world fighter. Even though the YF-17 lost the USAF competition, it was the prototype for the U.S. Navy's F/A-18 aircraft.
Size: 8 1/2" x 11" • over 100 color and b/w photographs, line drawings • 64 pp.
ISBN: 0-88740-910-5 • soft cover • $19.95

GENERAL DYNAMICS F-111 AARDVARK. *Don Logan*. Covers all models from the F-111A through the F-111K. In addition to a "Roll Call" including histories of each individual aircraft, photos of over 90% of the aircraft are included. Nose art of the FB-111A is covered with 48 photos. One Hundred and twenty Unit and Campaign patches are included, along with appendices on External Differences, Avionics Systems, F-111 Specifications, a Unit and Tail Code Summary, Attrited Aircraft, and a list of the 113 Aircrew members who died flying the F-111.
Size: 8 1/2" x 11" • 860+ color photos, line drawings • 320 pp.
ISBN: 0-7643-0587-5 • hardcover • $59.95

NORTHROP'S T-38 TALON: A Pictorial History. *Don Logan*. This is the story of the most successful pilot training jet ever produced: Northrop T-38 Talon. The history of the aircraft is broken down by the roles it has played in over thirty years of service. All units flying the T-38, their markings, and paint schemes are covered in over 300 color photographs-including a chart of colors used listing Federal Standard (RS) color numbers.
Size: 8 1/2" x 11" • over 300 color photographs • 152 pp.
ISBN: 0-88740-800-1 • soft cover • $24.95

ROCKWELL B-1B: SAC's Last Bomber. *Don Logan*. Covers the complete history of the B-1 Lancer from its inception, through production, and operations with Strategic Air Command. Tables are included which cover B-1 program milestones, communication and navigation equipment, aircraft specifications and performance, production number and serial number list, aircraft camouflage and paint schemes, and Aircraft name/nose art history. Photos of each configuration, paint scheme, and nose art of every B-1B are also included.
Size: 8 1/2" x 11" • over 400 color and b/w photographs, line drawings • 256 pp.
ISBN: 0-88740-666-1 • hard cover • $49.95

ACC BOMBER TRIAD: The B-52s, B-1s, and B-2s of Air Combat Command. *Don Logan*. The USAF has the singular ability to project power rapidly, precisely, and globally using strategic bombers. Air Combat Command's (ACC's) heavy bombers, the B-52H, B-1B, and B-2A are at the core of the USAF's provide both nuclear and conventional deterrent options for the United States. This book presents a roll call of ACC's bombers with photographs and history of all 208 bombers since joining ACC in June 1992. Also included are all of the bombers' weapons. Includes 80 unit and special purpose flight suit patches.
Size: 8 1/2" x 11" • 780 color photos, 40 line drawings • 304 pp.
ISBN: 0-7643-0680-4 • hard cover • $59.95

THE 388TH TFW AT KORAT ROYAL THAI AIR FORCE BASE 1972. *Don Logan*. Covers the 388th TFW; a Composite Wing based at Korat RTAFB, Thailand, consisting of fighters, Wild Weasel aircraft, airborne jamming aircraft and AWACS aircraft. Discusses in detail the Wing, the Squadrons and the aircraft they flew: the F-4. F-105G Wild Weasel, A-7D, EB-66, EC-121, and C-130. Also covered are the mission types, as well as operations of the Wing during the Linebacker Campaign over North Vietnam. Narratives of all the 388th MiG kills and aircraft losses during 1972 are included. A selection of official and unofficial flight suit patches is also included.
Size: 8 1/2" x 11" • over 170 photos • 128 pp.
ISBN: 0-88740-798-6 • hard cover • $29.95